George Bates Nichols Tower

Guide Posts on the Engineer's Journey

Ill. by Numerous Engravings and cContaining many Useful Tables

George Bates Nichols Tower

Guide Posts on the Engineer's Journey

Ill. by Numerous Engravings and cContaining many Useful Tables

ISBN/EAN: 9783337077082

Printed in Europe, USA, Canada, Australia, Japan

Cover: Foto ©berggeist007 / pixelio.de

More available books at **www.hansebooks.com**

GUIDE POSTS

ON THE

ENGINEER'S JOURNEY.

ILLUSTRATED BY NUMEROUS ENGRAVINGS

AND CONTAINING MANY

USEFUL TABLES

EDITED BY

GEO. B. N. TOWER,

CONSULTING ENGINEER; FORMERLY CHIEF ENGINEER UNITED STATES NAVY; INSTRUCTOR IN
ENGINEERING AND MECHANICS, CHANDLER SCIENTIFIC DEPARTMENT, DARTMOUTH
COLLEGE; AND LATE UNITED STATES SUPERVISING INSPECTOR
OF STEAM VESSELS, SECOND DISTRICT.

PUBLISHED BY
AMERICAN STEAM BOILER INSURANCE CO. OF NEW YORK.
1889.

List of Books and Authors Consulted.

PERMANENT WAY OF EUROPEAN RAILWAYS. . . .
      ZERAH COLBURN AND A. L. HOLLEY.

TREATISE ON STEAM BOILER INCRUSTATIONS. CHAS. T. DAVIS.

TREATISE ON STEAM BOILERS. . . . T. WILSON.

STEAM MAKING. . . . . PROF. CHAS. A. SMITH.

STEAM BOILER EXPLOSIONS. . PROF. R. H. THURSTON.

RAPPORT DE L'ASSOCIATION POUR LA SURVEILLANCE DES CHAUDIERES A VAPEUR. . . . . BRUSSELS.

RESEARCHES IN STEAM ENGINEERING. . .
      B. F. ISHERWOOD, U. S. N.

MARINE BOILERS. . . . . N. P. BURGH.

STEAM ENGINE. . . . . PROF. RANKINE.

"AMERICAN ENGINEER." . . . . CHICAGO.

PREFACE

This is largely a compilation from every reliable source bearing upon the subjects treated, as it would be almost an impossibility to write an entirely original work regarding them. It is intended simply as a book that can be carried in the pocket for a reference, as its many practical tables can not fail to be of some use, and the want of them is often vexatious.

In the chapter on fuels, some accounts of the various methods of using "petroleum" and "natural gas" are given, with illustrations of the apparatus employed, which it is believed are new to the general public.

A list of the books and authors more generally consulted and quoted from is inserted.

TABLE OF CONTENTS.

	PAGES.
Boilers, . . .	7
Method of Designing Boilers, ...	29
Specification for a Boiler, ...	33
Mode of Superintending the Construction of a Boiler, . .	37
Setting of a Boiler, ...	38
Boiler Management, . .	42
Boiler Fixtures and Appendages, . . .	47
Defects in Boilers, .	57
Explosion and Ruptures, . .	59
Fuel for Boilers, ...	63
Combustion, . . .	91
Incrustation, ...	106
Corrosion, ...	114
Method of Laying-out and Building a Boiler in the Shop, .	116
Miscellaneous, ...	125
Tables, ...	129

ERRATA.

Page 13.—Last paragraph. There is a difference of opinion regarding the method of carrying the flue over the top of boilers. Some engineers approve while others disapprove of this plan.

Page 15.—Last line, third word, "double," should be omitted.

Page 29.—Seventh line from bottom, first words should be "wrought iron," not "cast iron."

Page 35.—Fourteenth line from bottom, in paragraph "Heater," the second "one" should be left out; should read : "One exhaust heater to be provided and erected in place."

Page 38.—Eighth line from top. It is a matter of opinion as to which is the best, straight or circular bridge walls. The straight is oftener preferred.

Page 39.—Sixth paragraph. See errata for page 13.

Page 54.—Sixth line from top, "height of one (1) inch above highest row of tubes" should be "height of two and one half (2½) to three (3) inches above highest row of tubes."

Page 55.—Fourth paragraph, "Steam Domes." There is a difference of opinion regarding steam domes. Some engineers approve of their use.

Page 58.—Sixth line from top, near end, should read : "If a buck stave is broken."

Page 91.—Third line from top, "heated" should read "heat."

Page 100.—Eleventh line from bottom, third word from end, "is" should read "by."

Page 118.—Example should read :
6 × 6 = 36 (6 squared); 36 × .7854 = 28.2744 square inches.

$$\begin{array}{r} 36 \\ \hline 47124 \\ 23562 \\ \hline 28.2744 \end{array}$$

Page 124.—Boiler plate tests should read :
Tensile Strain of Iron per sq. in. } Lengthwise of the fibre 45,000 lbs. Crosswise of the fibre 40,000 lbs.
Tensile strength of steel boiler plate is from 50,000 to 70,000 pounds.

Page 125.—Eleventh line from bottom, 2¼ should be 2½.

Page 128.—Third paragraph, the caption should read : "To Ascertain the Bursting Pressure of a Boiler. For safe working pressure divide by six. If the boiler is double riveted add 20 per cent."

There are a number of typographical errors which the reader will observe, but space would not admit of noting.

Digitized by the Internet Archive
in 2008 with funding from
Microsoft Corporation

http://www.archive.org/details/guidepostsonen

STEAM BOILERS.

This is not a new subject, but it most certainly is a topic that ought to be of general interest to the whole community, as well as of vital importance to manufacturers and engineers.

What is a steam boiler, and what is its use?

Generally speaking, it is a closed metallic vessel, both strong and tight, in which steam is generated from water by the application of heat, for the purpose of giving motion to machinery, or of supplying heat in a more convenient manner to places where it is needed.

But a boiler is not complete without certain fixtures, appendages and accessories. There must be a feed pump, or injector, with a supply pipe, feed pipe, feed valve, safety-feed valve, and check valve, in order to supply water properly to the boiler; gauge cocks and a glass water-gauge to show the height of the water in the boiler, or *water-level*, as it is more commonly called; a heater, to assimilate the temperature of the feed water as nearly as possible to that of the boiler water; a blow-pipe, with its valve, to reduce the height of the water in the boiler, or to empty it entirely; a safety-valve, to allow the steam to escape from the boiler when it exceeds a fixed pressure, in order to prevent strains or rupture; a scumming apparatus, to remove the foreign matters from the water as much as possible; a steam pipe, to convey the steam to the place where it is wanted; a reverse valve, to prevent the formation of a vacuum in the boiler, and thus to avoid collapsing strains; man-holes and hand-holes, with their covers and guards, for examination and cleaning; a steam-gauge to indicate at all times the amount of pressure in the boiler; and a fusible plug to give warning in case of "low water," though this last is not absolutely a necessity. Also there should be braces and stays to strengthen the shell and prevent alterations in its form, as well as reinforce plates and hoops, and rings in the furnace flues of some styles of boilers.

All these are attachments to the boiler proper, having direct reference to its internal functions; but, in addition, there are the lugs, pedestals, or brackets, which support the boiler; the masonry in which it is set, with its binders, rods and wall plates; the boiler front, with its doors, anchor bolts, etc.; the arch plates, bearer bars, grate bars, and dampers, and last, but not least, the chimney and its topping. These are all equally necessary to enable the boiler to perform its duty properly. And, besides, there are required fire tools, flue brushes and scrapers, and scaling tools, with hose also to wash out the boiler, to say nothing of hammers, chisels, wrenches, etc.

Thus we see that in speaking of a boiler, not only the boiler proper is meant, but also the whole of its fixtures, appendages and belongings, making quite an assemblage of different parts and pieces, each and all of which must be kept in good condition at all times, in order to insure safety, and obtain the best results possible as regards the generation of steam.

Now, *What is steam?*

We might say steam is a vapor formed from water, but that is not sufficiently definite, and demands some explanation.

The passage of any liquid into the gaseous state is called *vaporization*, the term *evaporation* especially refers to the slow production of vapor at the free surface of a liquid, and *boiling* to its rapid production in the mass of the liquid itself. The term *vapor* is confined to *evaporation* without *boiling*, or *ebullition;* the term *steam* indicates the gaseous form of water produced by *ebullition*, which is commonly understood to take place at a temperature of 212° Fahr., or above it.

The experiments of Arago, Dulong, Regnault, and others, have long since determined that the boiling point of fresh water at the sea level, under the pressure of the atmosphere, corresponding to

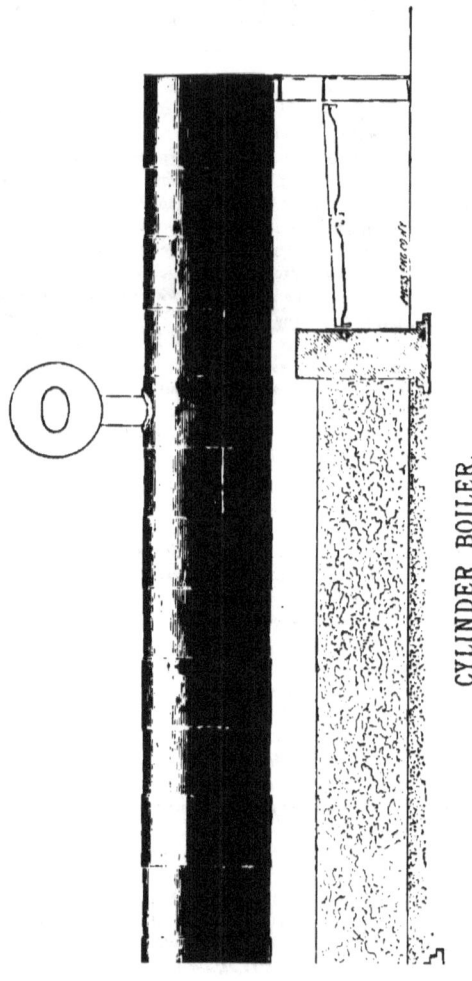

CYLINDER BOILER.

a height of 29.92 inches of the mercurial barometer, is 212° Fahr., or 100° centigrade, and also that, if the water is confined in a vessel under pressure, the temperature of the boiling point of the water rises when the pressure is increased, though not in the same ratio; but *the same amount of pressure always corresponds to the same temperature of the boiling point in the same liquid.*

And, again, if the water is not pure, or nearly so, but holds various salts in solution, the temperature of the boiling point is thereby increased for the same pressure. For instance, the boiling point of sea water, which contains salts of lime, magnesia, and sodium in solution, is 213.4° Fahr. under the pressure of the atmosphere. It will be well also to recollect that on the top of a high mountain the pressure of the atmosphere is less than at the level of the sea, and of course there will be less pressure on the surface of a body of water there, and consequently it will boil at a lower temperature, and its steam will have less tension.

A liquid boils when the tension of its vapor is equal to the pressure it suppports.

Advantage is taken of this fact in the vacuum pans of the sugar refineries, where the syrup is inclosed in tight pans, heated by steam coils, and evaporated at a low heat in a partial vacuum which is maintained by pumps.

The particles of water are strongly cohesive, but the particles of steam are repellant. It takes a certain amount of coal to raise the temperature of a cubic foot of water from 60° Fahr. to the boiling point—212° Fahr., but to further raise that water into steam of the same identical temperature would require a still further expenditure of coal.

Now, *what becomes of the extra amount of heat developed by this additional amount of coal?* It is not shown by the thermometer; it is absorbed in driving apart the particles of water, and keeping them apart in the gaseous state as steam, and it is called "latent heat" while thus employed; but, when the steam is condensed, it manifests itself by increasing the temperature of the water used for its condensation.

The amount of heat thus absorbed varies with the temperature, and has been very thoroughly nvestigated by Regnault, who has tabulated his results, completely overthrowing the old idea that the sum of the sensible and latent heats was always a constant at all pressures.

In order to generate steam in a boiler, it is necessary to fill it with water to the proper height, then to start a fire in the furnace to produce ebullition of the water, and to have the safety-valve weighted to the tension required of the steam. That all seems easy enough to do, and it really is so, when you know just how to do it, and not otherwise. But before discussing the management of boilers it is better to mention a few of the different styles.

Boilers may be divided into three great divisions or classes; those having internal furnaces, those which are externally fired, and those having detached furnaces, or ovens; and, again, each of these classes comprise two divisions, those in which the heated gases pass from the furnace through flues, or tubes, surrounded with water, on their way to the chimney; and those in which the gases pass between and among tubes filled with water. The latter are called "water-tube" boilers. There are also various patented boilers composed almost entirely of tubes filled with water, which are generally known by the names of the patentees, but they are classed as tubulous boilers.

The kinds more generally used in this country are "shell" boilers, as the "fire" and "water-tube" boilers are often called to distinguish them from "tubulous" or "pipe" boilers and they are:—

The Plain Cylinder; Flue; Drop Flue; Return Tubular; Double Cylinder, Union, Sullivan; Locomotive or Fire Box; Cornish; Lancashire; Galloway; Upright Tubular; Hog-Nosed.

Besides these there are many patented "shell" boilers, and some of them possess considerable merit,

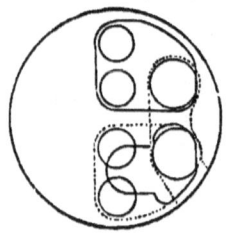

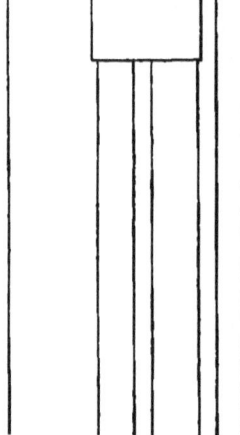
RETURN DROP FLUE BOILER.

GUIDE POSTS ON THE ENGINEER'S JOURNEY. 11

THE PLAIN CYLINDER BOILER is, as its name denotes, a simple cylinder composed of iron plates; usually it has flat heads each made from a single plate of iron, but sometimes these heads are made of cast iron, and at others, though made of plate iron, they are either dished or convex. The manholes of these boilers are generally made in the heads. It is more usual to set these boilers in "nests," or "batteries," as they are indifferently called, of two or more, in which case they are placed between two walls of masonry, and supported by hangers from cross bearers, usually resting on the tops of the walls which extend upwards above the shell of the boiler; but sometimes they are supported by the cast-iron fronts and standards, or by the mud drums. The furnace is common to all the boilers in the battery, extending from wall to wall athwartships under the front ends of the boilers, and at the back end of the grate bars is the "bridge wall." Under the farthest end of the boilers is another "bridge," and the space between these two bridges is often filled in to within a short distance of the tops of these bridges, and floored over with brick, forming a combustion chamber, or "flame bed," as it is more generally called. (See plate.) These boilers vary in length from 24 ft. up to 70 ft., and are usually from 30 inches to 42 inches in diameter, and generally no braces are required. In the Eastern States the usual mode of setting is without a mud drum, but west of the Alleghanies a mud drum is considered indispensable. When in battery these boilers are usually connected to a horizontal steam drum, extending across their front ends, by short necks, and it is usual to have only one safety-valve for each battery, which is placed upon this drum. Where a mud drum is used, it is placed athwartship under the boilers of the battery, near their back end, the back bridge being dispensed with, iron pedestals, resting upon foundations, support the drum, which is connected to each boiler by a vertical neck, and it is usual to protect the drum proper by arching it over with brick. The feed pipe and blow pipe both connect with this drum, and it is fitted with a manhole for the purpose of cleaning and examination. When the boilers are long, two mud drums are frequently used, one at about the middle of the length of the boilers, and the other near the back end. These boilers are decidedly the most accessible for cleaning and repairs, and have an unimpeded circulation, but they have a large water body, and their heating surface is not so disposed as to obtain the greatest efficiency; and, again, where they are of excessive length in proportion to their diameter, they are liable to great wear from unequal contraction and expansion. We may consider them as durable, but not very economical in fuel; but the expense for repairs is a minimum, which is probably one reason why they were so generally used in manufactories until within about 25 or 30 years. Their first cost is a minimum, but they require a maximum of floor space. However, where the water is bad, their use, or that of the Flue boiler, is almost a necessity. They are seldom to be found nowadays except at iron works, and west of the Alleghanies. According to Prof. Thurston, it is an exceedingly safe boiler, but if it does explode, owing to extraordinary carelessness or ignorance, the consequences are usually exceptionally disastrous. And the facts fully warrant the above assertion.

THE FLUE BOILERS are usually cylinders with two flues of considerable size extending from head to head. They are set in masonry, like the Plain Cylinder Boiler, but the products of combustion, after passing along the boiler bottom to the rear end, return through the flues to the front, where they enter the "breeching" and are led into the chimney. These boilers are set singly, and also in "batteries," and in this latter case they often have more than two flues. It is not quite so easy to clean or repair these boilers as the Plain Cylinder variety; they are not so durable either, giving out principally in and around the flues; and their circulation is not so good, but they are more economical in the consumption of fuel. They do not require as much floor space as the Cylinder Boiler, and they have a smaller water body. They are seldom more than thirty feet in length, and their diameter is generally from forty to fifty inches. When set in

12 GUIDE POSTS ON THE ENGINEER'S JOURNEY.

batteries they are furnished with mud drums in like manner as the cylinder variety. This variety is the style most in use west of the Alleghanies. It is more liable to explosion than a plain cylinder boiler, and the effect would be fully as disastrous. Its mode of setting is shown in the plate.

THE DROP FLUE BOILER has either one or two short furnace flues in the front end, and from the upper part of the back end of the furnace flues lead to the upper portion of a chamber, called the "back connection," in the rear end of the boiler, from the lower part of which flues lead towards the front end of the boiler, into a chamber called the "front connection," from which is an opening into a flue, formed in the masonry, leading along the bottom and sides of the boiler to the chimney. This boiler is set in masonry, but does not require so much floor space as the ordinary Flue Boiler, and is of the internally-fired class. The arrangement of the flues in this boiler permits of a long run for the gases, notwithstanding its comparatively short length, allowing sufficient time for their thorough combustion, and consequently its economy in fuel is a maximum, but its circulation is only fair, and it is not so accessible for cleaning and repairs as either of the two varieties already mentioned; its cost is greater, and it is neither so durable nor so strong, though steam can be carried steadily, owing to its large water body.

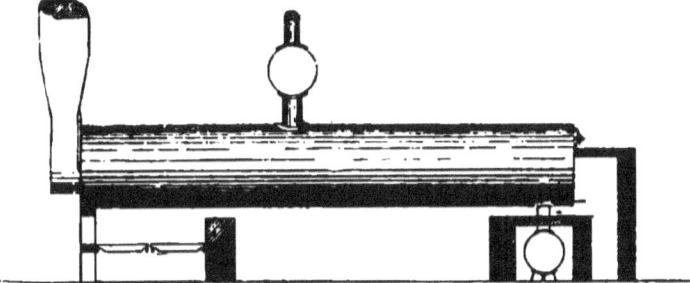

FLUE BOILER.

A HORIZONTAL OR RETURN TUBULAR BOILER has a cylindrical shell also, the lower half of which is filled with flue tubes extending from head to head. It is set in masonry, supported either by brackets riveted to its sides and resting upon the walls, or by the cast-iron front and a pedestal, and, of course, is externally fired, the products of combustion passing along the bottom and sides of the shell to the rear end of the boiler, where they enter the tubes and are delivered into the front connection at the furnace end of the boiler, and are led from thence into the chimney. Sometimes the gases are led from the front connection over the top of the boiler to its rear end, and thence direct to the chimney; this, however, is a very reprehensible plan, the practical results of which are to cause corrosion and wasting of the boiler top, as well as seam leaks, and it prevents access to that part of the shell for examination. As the tubes in this boiler are extinguishers of flame, it is highly necessary in order to insure the greatest efficiency, that provision be made in the furnace and combustion chamber for the complete combustion of the gases before they enter the tubes. It generates steam more rapidly than either of the boilers yet mentioned, and is nearly as economical in fuel as the Drop Flue variety, but its circulation is generally imperfect, while the facilities afforded for cleaning and repairs are very poor. Some of these boilers, however, recently designed, have an

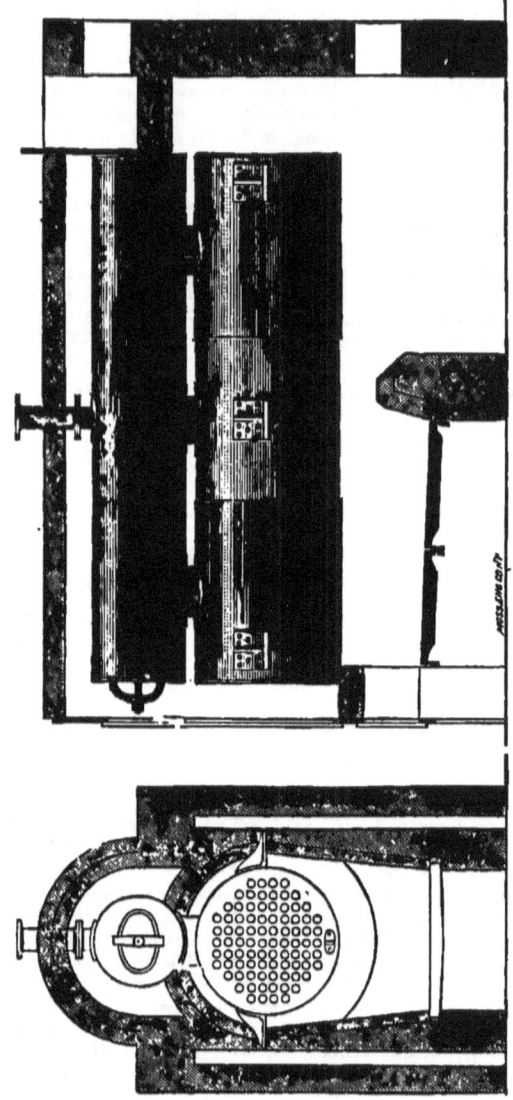

UNION BOILER

excellent circulation, are easy to clean and repair, and have obtained the maximum in economy of fuel. It is strong, durable, and occupies a medium of floor space, and is decidedly the best boiler for using Welsh or anthracite coal. It is a distinctively American boiler. When properly constructed and managed has small liability to explode, but when it does explode the damage is great.

THE DOUBLE CYLINDER, UNION, OR SULLIVAN BOILER.—These are the names of the different varieties of substantially the same boiler, the underlying principle being the same and the general appearance also, the variations being in non-essentials. It consists of two cylindrical boilers, one lying above the other, the lower cylinder generally being the larger, in which case it is filled with flue tubes extending from head to head. These cylinders are connected by short vertical necks. The upper cylinder has no tubes, serving principally as a steam

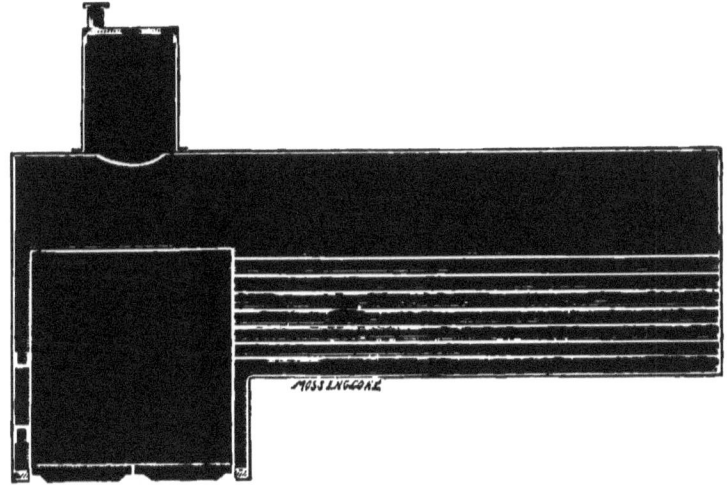

FIRE-BOX BOILER.

chamber, although a certain depth of water is maintained in it. Sometimes the lower cylinder is the smaller of the two, and in that case it is without tubes or flues. The first figure in the plate shows the general plan of setting of the varieties known as the Union and Sullivan boilers, and the second figure the Double Cylinder boiler proper, as used in the Western States. It is not an economical boiler in point of fuel or repairs. The Union and Sullivan varieties are not easily accessible for cleaning, etc , though the Double Cylinder variety is, but they are all constantly liable to rupture from unequal expansion at and around the connecting necks, while internal corrosion is usually developed in the necks themselves. The Franklin Institute strongly disapproves of their use. They are neither durable nor desirable, and, to sum up in a few words, they are *equal to few and superior to none.*

THE LOCOMOTIVE OR FIRE-BOX BOILER consists of a square or oblong box, with double double walls with arched top over the outer walls, and a nearly flat top over the inner

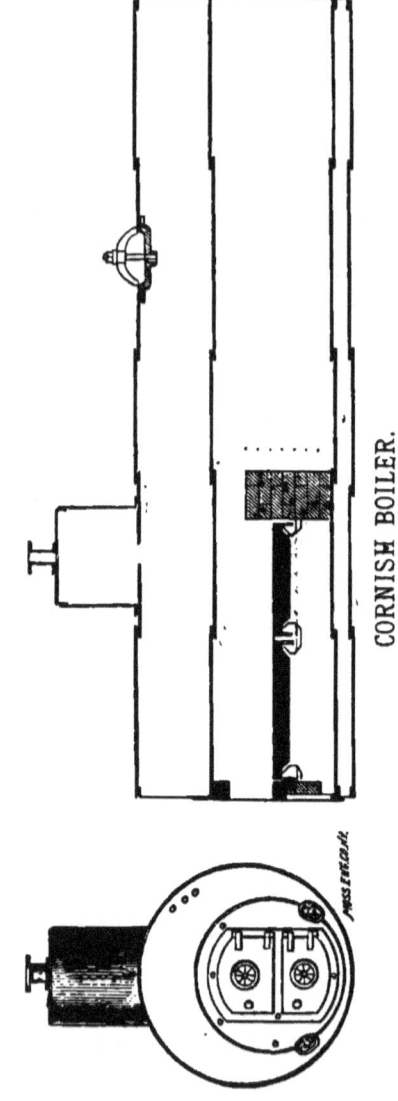

walls, which form the sides of the furnace, the bottom not being closed but urnished with a grate for the fuel. In one end of the box is the door for managing the fire. Into the outer wall, at the end opposite the door, a hollow cylinder called the "barrel" of the boiler, is inserted, which is furnished with a flat plate-iron head, and from the back end plate of the furnace flue tubes extend through to the head, and the gases generated in the furnace are led through them into the "smokebox," which is generally a prolongation of the barrel, furnished with a head in which is a door for examination of the tubes and the removal of dirt. From the top of the smoke-box rises either the chimney itself or a flue leading to it. The fire is entirely surrounded with water, and steam is generated very quickly, but it is not generally very economical in fuel, nor is it easily accessible for cleaning and repairs. It requires, however, no masonry, and can be easily mounted on wheels and transported from place to place. It is fairly strong and durable, but requires considerable care, especially on account of the small quantity of water it contains. It should never be used where the water is bad. It should always be protected as much as possible by felting or sheathing, to avoid loss by radiation.

THE CORNISH BOILER is cylindrical in form and has one large flue extending from head to head. The furnace is in one end of this flue, which is often four feet in diameter, with a bridge wall of brick at the back end of the grate bars. This boiler is usually set between parallel walls of masonry, and the gases, after passing out of the boiler flue, wind around and along the sides and under the bottom of the boiler on their way to the chimney through flues recessed into the brickwork. With slow combustion it is economical in fuel, but it occupies considerable floor space. When the boiler flue is constructed with Adamson expansion joints, properly spaced, it is very strong and durable. It is accessible for cleaning and repairs, and the circulation is good, but it does not generate steam quickly, but it holds the pressure well, as it has a large water body. It has not met with much favor in this country, as there are many other kinds much its superior. It is a distinctively English boiler. Its liability to explosion is much less than that of a Cylindrical Boiler, and its destructive effect is less. It belongs to the internally-fired class.

THE LANCASHIRE BOILER is also a horizontal cylindrical boiler, but having two large flues extending through it, with the furnaces in one end of the flues. It is set in brick work in the same manner as the Cornish Boiler, and it possesses about the same advantages and disadvantages, but it generates steam more quickly. It is really only a double-flued Cornish Boiler.

THE GALLOWAY BOILER is merely a Cornish or Lancashire Boiler, with Galloway tubes, which are simply conical water tubes, the larger end being uppermost; they are at intervals along the length of the flues, and across them, in both vertical and diagonal positions. These tubes promote circulation in the boiler, and add greatly to its heating surface and efficiency. It is very economical in fuel, is strong and durable, as well as fairly accessible for cleaning and repairs. When it has two furnace flues they generally open into a combustion chamber, and there is only one flue of a shape approaching an oval form with a convex top and slightly concave bottom to facilitate the setting of the tubes. Although these boilers require considerable floor space, yet they have been adopted for several large plants in this country, and so far have given satisfaction. They are not generally smaller than six or seven feet in diameter, and have a length of twenty feet. They are also English boilers.

THE UPRIGHT TUBULAR BOILER is an upright cylinder, having a concentric flat-topped furnace in its lower end. Tubes extend from the top of the furnace to the head of the boiler, and the gases pass through them into a "hood" covering the top of the boiler, and from thence generally into the chimney, which is usually a pipe of thin iron plate extending upwards from the hood. These boilers are strong, but not very durable, on account of their non-accessi-

GUIDE-POSTS ON THE ENGINEER'S JOURNEY.

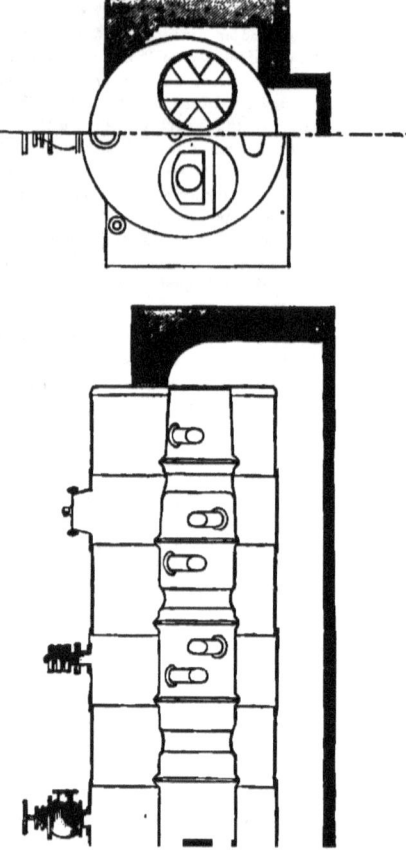

LANCASHIRE BOILER, WITH GALLOWAY TUBES.

bility for examination, cleaning or repairs; they require close attention, as they need repairs often. Their circulation is very defective, and they are wasteful of fuel. They generate steam quickly, as they have only a small water body, and they are often mounted on wheels, and are much used as portable boilers, more especially in connection with hoisting engines and drilling machinery, as also pumps. A boiler of this kind occupies but little space, and requires no masonry. When not n use they should be entirely freed from water, and their interior should be kept dry, to prevent corrosion and pitting of the tubes.

THE HOG-NOSED BOILER is a variety of the Locomotive Boiler. It consists o a cylinder having the lower half cut away for some feet at one end, and a flat plate, forming the top of the furnace, and a half head, forming the back end of the furnace, are used to cover the opening. From the half head flue tubes extend to the back head of the boiler and serve to carry the gases to the "back connection," from whence they are generally led to the chimney, although they are sometimes led towards the front, along the sides and bottom of the boiler, before passing to the chimney. This boiler is set in masonry, so that the "fire-box," or what corresponds to that part of a Locomotive Boiler, is made of brick, and consequently it belongs to the externally-fired class. It is not so strong and durable as many others; it is hardly fairly accessible for cleaning and repairs, but it generates steam quickly. It is not very economical in fuel, has a medium body of water, and needs care, as it often requires repairs, especially where the "hognose" joins the main body of the boiler. It requires only a medium of floor space; its circulation s poor, and its use cannot be recommended.

There are certain prime requisites for a good boiler, which are as follows:—

Strength; durability; small bulk and weight; free circulation of steam and water currents; efficiency of heating surface; facility of access for examination, cleaning and repairs; proper arrangements to secure prevention of smoke and thorough combustion, resulting in economy of fuel; and facilities for supplying the feed water at a temperature approximating closely to that of the boiler water, and as pure as possible.

It may be as well to remark here that much depends upon the location of the boiler. On shipboard there is little or no choice in the matter, but on land the circumstances are very different as there are not the same kind of limitations as to space; yet this very important point is often neglected until the very last moment, and even then, instead of a proper place being selected, it is put where there is hardly room enough to crowd it in. Now, what is the result? Sometimes it is so placed that there is no way of gaining access to the manhole, so that the boiler never can be examined or properly cleaned; or, it may be in a dark and damp cellar where corrosion is sure to ensue. It seems to be forgotten that the boiler is the life of a manufactory, and that when that stops working everything must come to a standstill.

The boiler should be placed on the ground level, or a oot or two above it, to insure drainage, in a light, dry, roomy place, and every facility should be afforded for keeping both the boiler and everything around it perfectly clean. The extra cost and trouble will be more than paid for in the resultant economy of coal and a reduced repair expense bill.

We will now notice some of the Tubulous Boilers in use to a greater or less extent in this country, and right here is the place to utterly deny and to pronounce as unqualifiedly false, the statements sometimes made by irresponsible and unscrupulous persons that boilers of this type are non-explosive. *That boiler never has been and never will be built that cannot be exploded.* And, furthermore, the stronger the boiler the more disastrous will be the explosion.

There are many varieties of this style of boiler, and some of them are undoubtedly familiar to many who may read this book; but it is well, nevertheless, to give them some consideration.

The same principle is common to all, viz.: —the heating of the water to a high degree quickly, by dividing its mass into many small portions. This is effected by having a moderate

UPRIGHT TUBULAR BOILER.

sized vessel or cylinder for a lower water body, and a much larger chamber for an upper water body and steam chamber combined, and connecting the two by inclined or vertical pipes, exposed to the heat of the furnace and gases, through which the mixture of water and steam rise into the upper water body, where the steam is disengaged, and the water generally flows back through vertical pipes, exposed only to the coolest gases, into the lower water body again.

These boilers must, of course, be set in masonry. Their principal differences consist in the arrangements of the tubes and water bodies, the style and methods of making joints and connections, and the positions of baffle plates and bridges to deflect the heat and gas currents.

In many places, more especially in mountainous districts where the only means of transportation is on the backs of mules, this style of boiler is a necessity, as it can be carried to its place of destination in easily handled parts, and put together on the spot. Generally every part of these boilers is accessible for cleaning or repairs, although it is frequently a work of time, owing to the many joints that must be broken and made over again.

The repairs required are generally few and infrequent, so that by keeping some spare parts on hand, they can be readily executed by a good engineer, without sending miles away for mechanics, and wasting valuable time. They are generally strong and durable; their circulation is not always good, varying with the inclination of the water-tubes; but their water body is small, and, consequently, they require close attention; also, they are not so economical in fuel as might be desirable, notwithstanding their claims to that effect.

But steam can be raised in them very rapidly, and expansion and contraction have a less injurious effect upon them, as a general thing, than upon other classes of boilers. Though not proof against explosions, the percentage of such accidents is a little less with this type than with the generality of boilers—though they are much more subject to ruptures, such as tube splitting, cracking, and blowing off caps at the joints. But one great defect is that the steam is rarely dry in this class of boiler, and, therefore, their use is not to be recommended.

THE TRIPLE-DRAFT, OR HENNESSEY BOILER, is a variety of the Horizontal Tubular Boiler. It is externally fired, and is set in masonry. A short cylinder projects from the lower part of the back head of the boiler proper, to which it is riveted. This cylinder is of considerable size, but is of smaller diameter than the shell of the boiler proper, and it is kept below the water line. Flue tubes extend from the head of this cylinder to the front boiler head, and also from head to head of the boiler proper, on both sides of the cylinder. The furnace is arranged as is usual under an ordinary Horizontal Tubular Boiler, but the short cylinder extends across the back connection, and through the usual back wall of the boiler setting, into a brick flue. The gases pass over the usual bridge wall under the bottom, and along the sides of the boiler into the back connection, thence back to the front connection through the tube of the boiler proper, and thence backwards into the brick flue through the tubes of the cylinder. It is strong, durable, and economical in fuel, but it occupies considerable floor space. An illustration of this boiler appears on page 24.

SLOANE'S STEAM GENERATOR —This is one of the numerous contrivances designed to utilize more fully boiler furnace heat. It consists of lines of suitably strong tubing arranged around the inside of the furnace well in such a manner as to protect its surface from the heat, which it utilizes. The tubes have an upward inclination of half an inch to the foot to a receiving chamber that is in turn connected with the upper part of the boiler, from the lower part of which they draw a supply of water. A convenient blow-off is also part of the apparatus. The exposure of the water to the fiercest heat of the fire enables the apparatus to return it to the boiler at steam temperature. The generator is claimed to add 200 to 250 square feet of heating surface to a boiler, and is used on old or new boilers that are not equal in capacity to the demands made upon them.

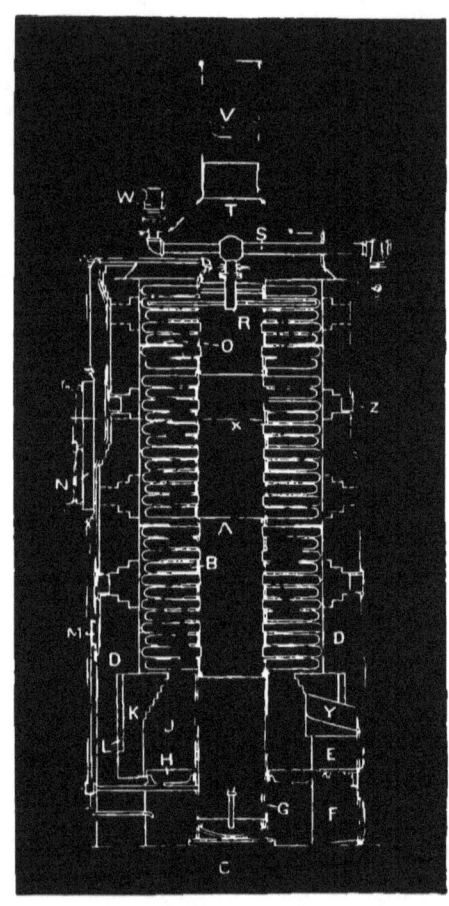

THE HAZELTON BOILER.
SECTIONAL ELEVATION.

The economy of this device will at once be apparent: it takes no heat from the boiler that would otherwise prove effective; it requires only what the boiler cannot use, and is so constructed as to turn it to good account. The deposit of sediment, etc., either in the generator or boiler, as well as the formation of scale, is rendered next to impossible by the rapid upward circulation, and such sediment as may be accumulated is easily blown out through a suitably arranged cock.

This apparatus has been used for more than eight years in factories, etc., and has proven itself to be reliable, and it is claimed that from thirty to fifty per cent of fuel is saved by its use. The steaming capacity of the boiler is increased some forty or fifty per cent at the same time, and on account of its supplying the boiler with water of such a high temperature the damage from expansion and contraction is much less, thereby increasing ths longevity of the boiler.

THE HAZELTON OR PORCUPINE BOILER is an original type of water-tube boiler, embodying principles sufficiently novel to merit special description, and is likely to become generally popular. Illustrations of its cross and vertical sections afford an excellent idea of the arrangement of its parts. A is the stand-pipe, into which are screwed or securely expanded the radial tubes B, the foundation C, and brickwork inclosure D, representing the masonry walls of the ordinary boiler. E is the furnace door, F the ashpit door, G the manhole, I' the furnace door arch, Z sight-doors, H the grate bars, and J the deflecting urnace, of which K' forms the fire-brick deflector, and L the air space insulating the latter from the brickwork inclosure. The fittings differ little, except in location, from those of any first-class steam-boiler, M being the steam-gauge, N the water column, S the steam-pipe with valve, T the smoke-hood, V the smoke-stack, W safety-valve and X the water-line. At O deflecting plates are placed, and R is a steam-drying device.

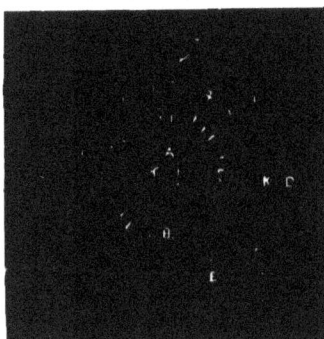

THE HAZELTON BOILER.
PLAN.

The advantages of this form of boiler are found to be various, and are well worthy the attention of steam users. The peculiar construction of the deflecting furnace insures thorough combustion of all fuel products and the complete utilization of their calorific value, the currents of heat being, by its form and by the employment of the deflecting plates above, turned inwards to the stand-pipe, where the radial tubes are close together and the heating surface of a character to absorb all the caloric generated, insuring rapid and economical steaming and thorough circulation, and the effectiveness of all heating surface. The arrangement of the radial tubes, with one end free, provides for perfect freedom of expansion and contraction, while the free position of the boiler with regard to its brickwork inclosure prevents tendency to leak or strain from this cause.

A manhole at the lower end of the stand-pipe facilitates thorough examination of the interior of the boiler, while the sight-holes in the brickwork afford convenient means for the introduction of a steam nozzle in removing soot and dust deposits from the exterior. The brickwork forms a chimney which can be raised to any height demanded to secure natural draft, to carry smoke to a proper heighth, etc., by the addition of a light iron stack. Owing to the small size of its parts and their ready accessibility, this is considered an exceedingly safe and convenient boiler; the employment of no cast iron, except in the grate bars, also increases its durability. Where small floor space is available and high pressure is desired, the Hazelton boiler is said to prove highly effective and economical.

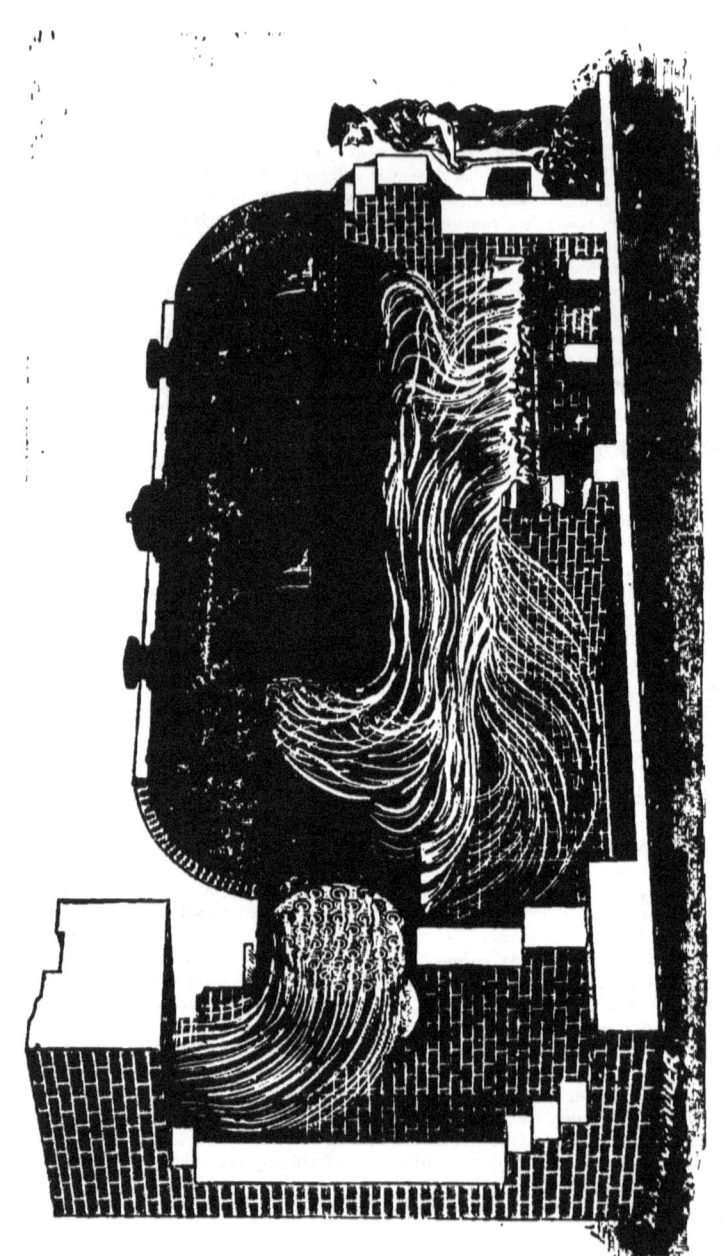

THE TRIPLE-DRAFT, OR MENNESSEY BOILER.

THE HEINE BOILER.—The Heine boiler, shown in the cut below, first came out in Berlin in 1879, and was introduced at St. Louis, Mo., in 1882. The heads of the shells are dished, and rectangular water legs, formed of a head and a tube sheet flanged towards each other and connected by a butt strap, are riveted near the ends of the shell to the lower third of the circumference. A nest of staggered tubes extends between the tube sheets of these water legs. There is a handhole plate in the head of each water leg, opposite the end of each tube. The tube sheets and heads of each water leg are stayed by large tubular stay bolts, so that a steam nozzle may be introduced for the purpose of freeing the tubes from deposit of ashes and soot, and plugs are loosely fitted which can be withdrawn easily to permit the passage of the steam nozzle.

The throat area of these water legs being practically equivalent to the total cross-sectional area of the tubes, insures a strong and uniform circulation of the water, and as this boiler is set at an inclination from the horizontal of 1 in 12, this also greatly promotes the circulation.

Two sets of tiling are arranged—one, resting upon the lower water tubes, extending from the front to within a short distance of the rear water leg; the other, resting upon the upper tubes, extending from the rear water leg to near the front—which compel the gases to traverse the tubes with an upward trend, and finally, after leaving the tubes, to travel once more under the boiler from front to rear, where they are conducted to the chimney.

All the heating surface can be easily inspected while the boiler is in operation by removing the loosely fitted plugs from the hollow stays and looking through them. Each boiler shell contains a submerged mud-drum in which the feed water travels the full length and returns before entering the circulation of the boiler proper, so that all mechanical and chemical impurities contained in the feed water are precipitated in the drum and blown out without entering the boiler itself. These boilers have a record of good economy, dry steam, and great capability of being forced whenever occasion demands. Both the plates and tubes of these boilers are made of the best flange and fire-box steel, no parts being constructed of cast iron except such small caps and plates as are universally allowed to be perfectly safe when made of this metal.

These boilers have the advantages of occupying a minimum of floor space with a maximum of efficiency; they are strong and durable, and are easily accessible for repairs; the steam space is large, and so is the area at the water level, which last necessarily conduces to the production of dry steam; while the arrangements for the supply of the feed water to the boiler are such that no contraction of plates can occur from the use of cold water, should it ever become necessary, thus almost entirely obviating the liability to grooving at roundabout seams on the boiler bottom; and also owing to the arrangement of the tiling there is little danger of the laps of the roundabout seams becoming turned or granulated.

There are over 800 of these boilers in use in this country and in Europe, and not a single case of explosion or serious rupture has yet occurred, nor has there been any damage to life or property caused by them since the first boiler was built in October, 1879.

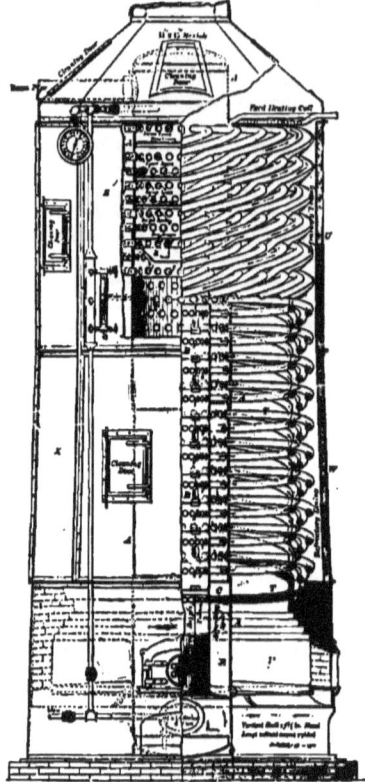

Fig. 1.
THE "CLIMAX" STEAM GENERATOR.

Fig. 2.

EXPLANATIONS.—A, Body of Boiler. B, Water Reservoir. C C, Supply Tubes. T T, Heating Tubes. S, Deflector. V, Fire-box. W W, Casing. Z, Refractory Lining.

THE CLIMAX STEAM GENERATOR.—This generator, invented by T. F. Morrin, has only lately been put upon the market, although it has been in successful use for four or five years at the tobacco works of P. Lorillard, and also upon the steam yacht *Reva*. In the accompanying cuts, Fig. 1 shows the apparatus partly in section and partly in elevation, while Fig. 2 is a horizontal section. Similar letters in both figures denote like parts.

It consists of a vertical cylinder, A, to the exterior of which a series of loop-tubes T, the upper and lower ends of which are not in the same vertical line. This cylinder is fitted with a manhole at its upper extremity, and there is also another in the shell near the lower extremity, below the line of the grate bars.

There is a smaller cylinder B inside of and concentric with the main cylinder, made in short sections which can easily be removed, if needed for repairs, its bottom being supported on brackets riveted to the outer cylinder, but its top is open and situated a trifle below the water line.

The lower ends, E, of the tubes, T, are connected to the inner cylinder by short tubes, C, crossing the space R, R. These tubes C are merely driven into the ends of the tubes T, and their other ends merely rest in the holes through the shell of cylinder B, not being expanded, as tight joints are not needed.

The fire-box surrounds the cylinder, A, the grate being annular, and its outer wall is formed of a thin iron casing lined with refractory material.

The feed water is heated by passing through a coil resting upon the upper tubes of the generator, and enters the cylinder B. The water and steam rise in the annular space, R R, and the tubes, T, while solid water descends in the cylinder B.

The deflector S serves to separate any water entrained by the steam. The diaphragm plates above the deflector cause the steam to pass through the upper tubes, thus drying and super-heating it.

The ratio of heating surface to grate area is large, being 50 to 1, and the circulation is unexcelled. This boiler is very durable, efficient, and economical in fuel; it occupies minimum of floor space, and requires little in the way of repairs.

It is unquestionably one of the strongest and safest boilers in use, and is especially adapted for high pressures.

ECONOMY IN THE USE OF EXHAUST STEAM.

Architects, Steam-Fitters, Engineers and Steam-Users have long been interested in the advantages to be derived from the utilization of the exhaust steam from engines or steam pumps, instead of wasting it into the air. By employing it as a source of heat to warm manufactories, office buildings, apartment houses, etc., and for certain manufacturing processes requiring heat, a large part of the percentage of the thermal value of the fuel consumed in a boiler furnace, which is ordinarily lost, is utilized, thus ensuring economy in fuel.

The exhaust steam in its normal condition, as it comes from the engine cylinder at a temperature of a little more than 212° Fahr., can be and is used to some advantage and economy under certain conditions; but practical experience has shown that, owing to its low temperature, moist condition, and comparative slow velocity, it is less efficient for the purpose of conveying and radiating heat than steam of a higher temperature; that it is liable to sudden and rapid condensation; and that it is difficult to obtain a free circulation for heating purposes, which eventuates in back pressure on the engine.

For the various manufacturing processes requiring a temperature of steam above 212° Fahr., in order to impart the necessary heat to the appliances in which it is used, exhaust steam in its normal condition is entirely unavailable without a degree of compression and consequent back pressure, which would fully neutralize its value—by reason of the extra fuel required to enable the engine to carry the additional load—and thus destroy any economical results from its use.

It has been found that while many buildings are supposed by their owners to be heated by exhaust steam only, yet that really during cold weather a considerable volume of live steam, to supplement the exhaust, is injected into the heating pipes, requiring the consumption of a large additional amount of fuel; or else that a back pressure of from six (6) to ten (10) pounds of steam is being carried on the engines or pumps in order to force the exhaust steam into circulation.

It must be evident to any one that if from 100 to 125 degrees of additional temperature can be imparted to the exhaust steam, thus raising it from its normal temperature of about 212° Fahr., up to say 300 or 325 degrees, which necessarily re-evaporates its contained moisture of condensation and expands its volume, both its capacity for radiating heat, and the facility with which it can be circulated through the heating system must be immensely increased.

Now if this result can be accomplished without extra expenditure of fuel, by means of heat that would otherwise be going to waste, and without impairing the draught of the furnace fires, the economy resulting therefrom cannot be disputed.

The temperature of the gases escaping into the chimney of a steam boiler vary from 400 to 600 degrees; this waste heat, and that of the exhaust blowing into the air combined, amount to 75 per cent of the thermal value of the fuel, which is utterly lost in an ordinary steam plant. A large percentage of this hitherto lost energy of the fuel can be recovered and brought into use by employing the waste heat or the purpose of imparting 100 or more degrees of added temperature to the previously wasted exhaust steam, and then bringing this steam into effective service for heating purposes.

The Hussey Reheater effects the above-named results, and is the greatest advance in economy of fuel yet made in steam-heating apparatus.

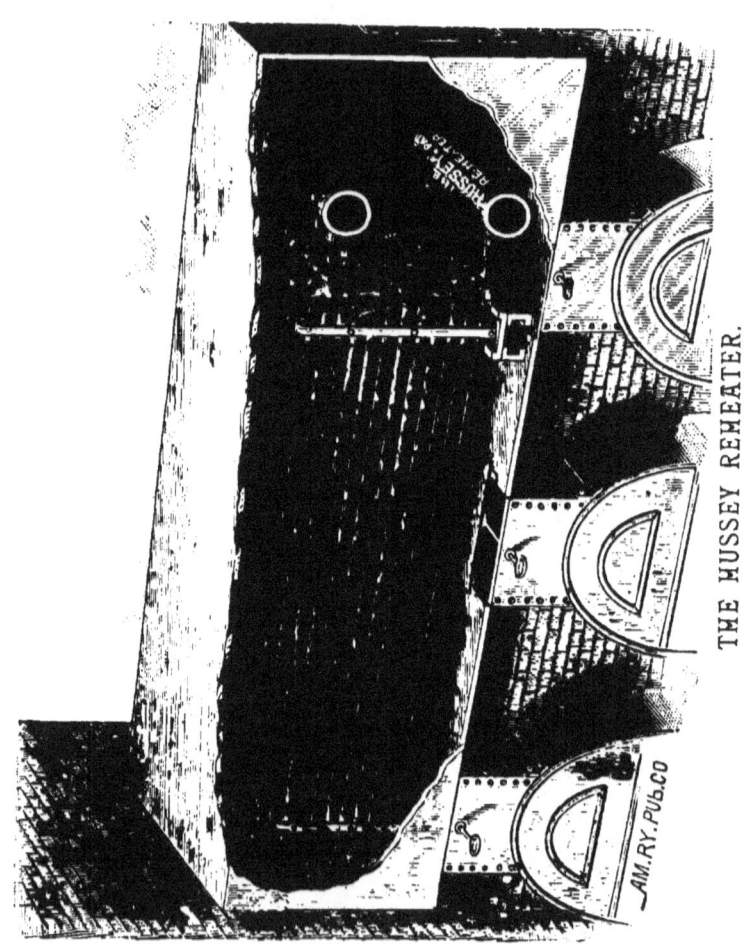

THE HUSSEY REHEATER.

METHOD OF DESIGNING A BOILER.

The great trouble at the present day in many plants is due to the fact that more is required rom the boilers than they were originally designed to perform; and there is also another trouble, due to the estimation of the capacity of a boiler by horse powers, which is sheer nonsense. A boiler should be estimated by its evaporation: with so much coal per hour it will evaporate so much water, from a given temperature, into steam, at a given pressure; with a different amount of coal, it will evaporate so much more or less, as the case may be.

An engine develops horse-power from the steam which has been evaporated from water by the boiler. With one engine, perhaps, 50-horse-power may be developed while the boiler is consuming a certain amount of coal; but, if that engine be removed and another be substituted for it, one of two things will probably happen; either the engine will develop 50-horse-power, while the boiler consumes less coal, or, on the other hand, while the boiler consumes the same amount of coal, the engine develops 25 per cent more horse-power.

Now these things are matters seldom known to the owners of the steam plants, though there is hardly a man who even pretends to be an engineer that is ignorant of these facts.

Boilers ought to be designed for the work they have to do, and the person to design them should be an engineer, not the boiler-maker, for his proper business is only to do the work in the best possible manner, according to the plans of the engineer.

Perhaps a few remarks as to the proportions to be observed in making the plans for a boiler may not be amiss, as in designing a boiler several things require close and careful consideration and attention.

In the first place the duty required of the boiler must be ascertained and defined in terms of pounds of water evaporated per hour.

Now, it is well known that when a boiler is new, with all its surfaces clean, it will evaporate more water with a given amount of coal than at any time subsequently, when its surfaces have become covered with scale and soot; so that if it will do its work easily when new, it will do it only with difficulty, or perhaps even fall short of doing it, after it has been in use a short time. Therefore, to insure a boiler's efficiency, it should be made larger, of greater power than is demanded in daily use, and it has been found that from 20 to 40 per cent is a proper allowance, dependent upon varying circumstances.

The author's practice has been to allow 25 per cent generally, under ordinary circumstances; but where the business is of such a nature that an increased duty is liable to be required from the boiler on account of adding new machines to the working plant, etc., 40 per cent is none too great.

Again, the material of which the boiler is to be made must be determined upon, whether it is to be of iron or steel. This is a matter in which the designer can only offer his advice, with his reasons therefor; the person for whom the boiler is to be built generally deciding that point himself. One thing, however, must be insisted upon by the designer, which is that only one kind of material shall be used in the shell. If steel is the material chosen, nothing but steel plates must be used in the construction of the shell; and the same precaution must be observed if cast iron is the metal selected. The two metals differ in their rates of contraction and expansion, and, consequently, by making use of them both in the same structure, an element of weakness and destruction is introduced. Furthermore, the quality of the iron, if that be the material selected, must all be of the same grade and from the same mill. And now let me say that the heads of boilers, as a rule, are made of too light material, and also that the flanges are frequently turned with too short a radius, which last should be carefully avoided.

It is next in order to determine the style of the boiler to be built, and the choice, unless the owner has a prejudice in favor of some particular kind, will generally depend upon four conditions, viz. :—

The kind of fuel intended to be used;
The quality of the water supply;
The pressure of steam required;
The space that can be allowed for a boiler room.

With Anthracite, Coke, Semi-bituminous Coal, or Natural Gas, there is no better boiler than one of the Return Tubular type; but with Bituminous Coal, Lignite, Wood, Sawdust, Oil or Residuum, a Flue Boiler is generally preferable.

Again, if the water to be used contains much foreign matter in solution, or is very muddy, so that considerable scale or sediment is formed in the boiler, either a Flue Boiler or a Return Tubular of the latest design is to be preferred, dependent upon the space afforded and the pressure to be carried; for, although a Plain Cylinder Boiler would give less trouble in keeping clean, it is too wasteful in fuel, as well as requiring a maximum of floor space.

As regards both high steam pressure and floor space, the Return Tubular Boiler is preferable.

Suppose, by way of illustration, that it is required to design a Return Tubular Boiler of 80 horse-power, so-called, to be built of steel, to use water that makes considerable scale, and to carry a steam pressure of 100 lbs. per square inch.

A horse-power, according to the engine in which the steam furnished by the boiler is used, may require the evaporation of from 15 to 60 pounds of water per hour; therefore we must first ascertain what kind of an engine is to be supplied with steam from the boiler. Let us suppose, in this case, that 30 pounds of water per hour evaporated in the boiler will supply the engine with steam enough to develop 1 horse-power = so that the water to be evaporated will be $80 \times 30 = 2,400$ lbs. per hour = to this we must add 25 per cent, and we shall have $2,400 + 600 = 3,000$ lbs. of water. One pound of good coal will evaporate, with good firing, 8 lbs. of water; consequently $3000 \div 8 = 375$ lbs. of coal required to be burned per hour. With a fair rate of combustion, $12\frac{1}{2}$ lbs. of coal will be consumed per square foot of grate per hour, and we shall have $375 \div 12\frac{1}{2} = 30$, which is the area of the grate in square feet; we should probably make this grate 6 feet wide by 5 feet long for convenience in firing. The area over the "bridge wall" should be about one-fifth of the grate area or 6 square feet, a ratio which has been determined by comparison of results of the best practice both in this country and in Europe. Also, it has been found in like manner that the heating surface of a boiler should be from 25 to 40 times the area of the grate. Now, suppose we assume 30 as the ratio for our boiler, and we shall have $30 \times 30 = 900$ square feet as the amount of heating surface required.

The heating surface in a Return Tubular Boiler consists of the surface of the lower half of the shell, the effective surface of the tubes, and the part of the back head below the water line, deducting the tube area. So now we have to determine the length and diameter of the boiler, and the number and diameter of the tubes.

From the results of the best practice it has been determined that from $\frac{1}{5}$ to $\frac{1}{7\frac{1}{4}}$ of the grate area is the proper proportion for the clear area through the tubes. The grate area has already been found to be 30 square feet, or 4,320 square inches, and $4,320 \div 7.25 = 595.86$ square inches, which is the total clear area through the tubes. Now, let us take 4 inches as the diameter of the tubes, which is very suitable for a boiler of this size, and as the thickness of a 4-inch tube is 0.13 inches, the inner diameter will be 3.74 inches, and the corresponding area will be 10.992 square inches, and we shall have $595.86 \div 10.992 = 54+$, so that we have 54 as the number of tubes required.

It now remains to determine first the diameter of the boiler, and afterwards its length. This boiler is to be a Return Tubular of the latest design, and it differs from the ordinary Return

Tubular principally in having a larger diameter, in order that there may be a manhole in the bottom part of the front head under the tubes, and also in arranging the tubes in banks separated by wide vertical spaces, so as to give great facility for cleaning and repairs, and also insuring a much freer circulation of the steam and water currents, which adds greatly to the efficiency of the boiler and promotes its longevity.

A boiler of this kind, properly set and well cared for, is very economical in fuel, while the cost of repairs is reduced to a minimum.

Now, in order to ascertain the diameter it will be necessary to make some rough calculations, and also to make use of a diagram, so that we can obtain the correct arrangement and position of the tubes and manhole, and provide that the clear spaces for water and steam circulating passages are of sufficient size.

It is well known that a square is almost one-fourth larger in area than a circle inscribed in it, now as the tubes in a boiler occupy only the lower half of it, if the boiler was filled with them there would be double the number of them—in this case 108—and if the boiler were square instead of circular in section, there would be one-quarter more of them, or $108 + 27 = 135$ tubes; but 135 is not a perfect square – the square root of it, however, is nearly 12—we will therefore take 12 as the number of tubes on the line of the horizontal diameter of the boiler. Now we will take a piece of paper ruled in five squares, say one-eighth of an inch on a side, and make a dot for a center at the corner where any four squares join, and draw a horizontal and also a vertical line through it for horizontal and vertical diameters of our boiler. We will take the tubes as divided into 4 banks of 3 tubes each, and will lay out 5 inches on the horizontal diameter for the clear space in the center, which will give 9 inches between the centers of the two middle tubes; again, we will set the tubes of a bank at a distance of one inch in the clear horizontally, which will make a distance of 5 inches between their centers. We have thus the positions of the tubes in the upper row of the two middle banks marked, and now we will lay off a space between the middle and outer banks of 4 inches in the clear, or 8 inches between the tube centers, and then lay off the centers of the remaining tubes of the outer banks on the horizontal diameter, spacing them as in the middle banks. Now, measuring on the horizontal diameter from the center to the outside of the farthest tube, we find that we have a distance of 2 feet 10½ inches, and if we allow 4½ inches of space between the tube and the shell, which is none too much, we shall have for the semi-diameter 3 feet 3 inches, or for the inside diameter of the large course of the boiler 6 feet and 6 inches. We will now strike a circle of that diameter from the center, and then rule vertical lines through the centers of the tubes already established. Now, starting from the center, we will divide the vertical diameter downwards into spaces of 5 inches, marking each division point, and then draw horizontal lines through these points, their intersections with the verticals already drawn are the centers for the tubes. Now as a manhole of 12×16 inches is to be laid out, we measure up from the inner edge of the shell on the vertical diameter for the lower edge of the manhole, and 6 inches above that we draw a horizontal line for the long diameter of the manhole, and proceed to draw in the manhole. We now finish drawing in the tubes, taking care that no tube shall approach the shell nearer than 4½ inches.

It now remains to find the length of the boiler, and to do this we must find the amount of heating surface in one foot of length of the shell and tubes, and then divide this amount into the total heating surface required, the quotient, of course, will give the length of the boiler.

The surface of a tube 4 inches in diameter and 1 foot long is 1.0472 square feet, therefore the total surface of 1 foot in length of all the tubes in this boiler will be $1.0472 \times 54 = 56.5488$ square feet. Generally we take only two-thirds of the surface of a tube as effective heating surface, and $56.5488 \times \frac{2}{3} = 37.6992$; but sometimes three-quarters of the tube surface is reckoned as effective, in which case we should have $56.5488 \times \frac{3}{4} = 42.4116$. In this case let us take a mean between the two, and we shall have 40.0554 square feet as the heating surface in 1 foot of length of the tubes. The surface of 1 foot in length of one-half the circumference of the shell, having a diameter of 6½ feet, is 10.2102 square feet; and the sum of all the heating surface of 1 foot in length is $40.0554 + 10.2102 = 50.2656$, and $900 + 50.2656 = 17.8$ feet for the length of the boiler, or 17½ feet, as the heating surface of the back head has not been taken into the account.

Now these dimensions, found as shown, will give a boiler that will always furnish steam enough to an engine for the development of 80 horse-power, provided the engine does not call for more than 30 lbs. of water to be evaporated per H.P. per hour; and it will not only do it easily, but can be urged to nearly 25 per cent more if it is absolutely necessary, without any great diminution of economy.

Now all that remains to be done is to calculate the thickness of the shell to stand the strain of a working pressure of 90 lbs. per square inch.

The table published by the U. S. Inspectors allows for a tensile strength of 60,000 lbs., when the horizontal seams are double riveted, and the plates are ⅜ of an inch thick, a pressure of 115 lbs. The heads should be $\frac{7}{8}$ of an inch thick.

In regard to the bracing, it should be so spaced that no brace should have a strain of more than 6,000 lbs. per square inch of section. It is a good plan to rivet strong bars of T iron to the heads, and secure one end of the braces to them, and the other end being riveted to the shell with two rivets.

The preliminary sketches and calculations having been now made, the designer can now proceed to draught the boiler to a suitable scale, taking a tracing therefrom to accompany the specifications, and retaining the original drawing.

After the drawing is finished and the tracing made, a specification should be drawn up to which the tracing should be firmly attached, and which should be distinctly mentioned as forming a part of the specification; and as this is for the purpose of obtaining estimates, any contract made must be based strictly upon the specification.

On the following pages is a form of specification for the 80 H. P. Boiler that we have just been figuring on.

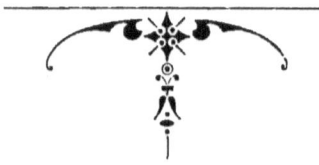

GUIDE POSTS ON THE ENGINEER'S JOURNEY. 33

SPECIFICATIONS FOR AN 80-HORSE-POWER BOILER.

STYLE.................Return Tubular.
MATERIAL.............Steel of 60,000 lbs. T. S. Thickness, ⅜ inch. To show a ductility of 50 per cent.
DIMENSIONS..........78 inches in diameter inside large course; 17½ feet in length from out to out of heads.
SHELL................To be in 3 courses, one sheet only to each course, and the middle course to be the small one. To be so set that the back end will be 1½ inches lower than the front.
HEADS................To be ⅝ inch in thickness, and the flanges to be turned on an inside radius of 2 inches, and to be free from cracks, flaws and deep hammer marks. The manhole flange in the front head to be planed to a smooth, even surface.
RIVETING.............All girth seams to be single riveted; all longitudinal seams to be double riveted and staggered. Rivets of shell seams to be ¾ inch in diameter. Rivets for flanges of heads to be ⅞ inch in diameter. Rivets for reinforce plates to be ⅞ inch in diameter. Rivet holes to be so punched as to come fair without the use of a drift-pin, and the holes are to be punched ⅛ inch smaller than full size, and then reamed to size. A spiral punch must be used for all rivet holes.
BRACES...............To be made of best stay bolt round iron of not less than 1¼ inches in diameter, each in one piece, and no brace to be less than three (3) feet in length. The foot of each brace to be secured to the shell by two (2) ⅞-inch rivets. The jaws of each brace to be secured to the T-iron brace bars on the heads by turned bolts, and the holes in both brace bars and jaws to be drilled. The brace bars to be secured to the heads by ⅞-inch rivets. Braces to be arranged as shown in accompanying tracing.
MANHOLES...........There must be one manhole in top of shell on the middle ring course, 12×16-inch opening, fitted complete with frame, yokes, bolts, nuts, plate, ring and gasket. There must be one manhole in lower part of front head, 12×16-inch opening, fitted complete with yokes, bolts, nuts, ring, plate and gasket. The flange of this manhole must be 1½ inches deep, and turned on a radius of ¾ of an inch. All these to be in accordance with the accompanying tracing.
HANDHOLE...........To have one handhole, 4½×7 inches in the back head, as shown in accompanying tracing, fitted with guard, bolt, nut and gasket.
NOZZLES.............To have two (2) nozzles of cast iron, with openings of five (5) inches in diameter, one on the top of the front ring course for the steampipe, and one on the top of the rear course for the safety-valve, as shown in the tracing, both to be riveted to the shell.
BRACKETS...........The shell of the boiler must be fitted with four (4) cast iron brackets for supports, two (2) on each side, as shown in the tracing. The front ones to rest upon iron wall plates ⅜ inch in thickness, the back ones upon three (3) iron rolls (double thick gas pipe) each, the

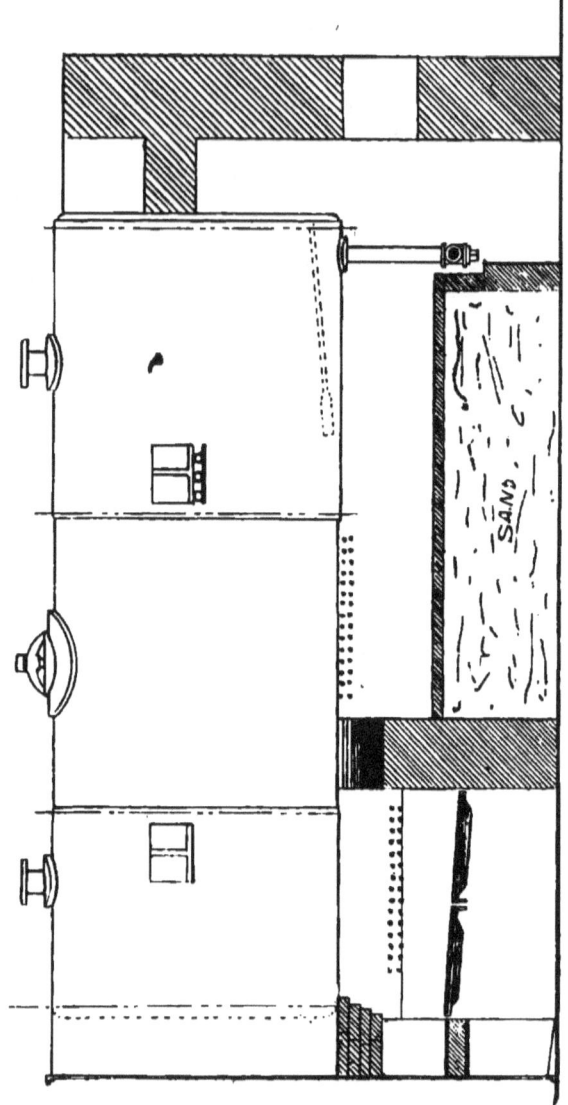

LONGITUDINAL SECTION.

GUIDE POSTS ON THE ENGINEER'S JOURNEY. 35

	rolls to have a diameter of 1 inch, and to be supported upon 1½-inch iron wall plates.
BLOW-OUT	A circular patch, ½ inch in thickness, to be riveted to the bottom of the boiler, near the back end, with ⅞-inch flush rivets, and to be drilled and tapped for a 2½-inch blow-out pipe.
FRONT	A cast iron front to be provided, and all doors for facility of access to tubes, furnaces, ashpits and air spaces. Furnace doors to be fitted with registers and perforated linings. Ashpit doors to have registers only. All bolts, cramps and anchor bolts to be provided.
BACK CONNECTION	A door, 20x16 inches, to be fitted into the back wall at end of boiler, or into the side wall back of the bridge, to allow of access to the back connection, and angle, or T-iron bars to be furnished to support the covering of the back connection.
DAMPER	A damper to be fitted in the connection to the chimney, with suitable means for operating it.
BUCK STAVES	To be provided with six (6) buck ▇▇▇, three (3) on each side, with all bolts, nuts, washers, p▇▇ ▇nd anchor bolts.
HOLES	All holes for pipes, gauges, fittin▇▇ ▇tc., to be drilled and tapped as may be required.
FITTINGS	1 3½-inch Safety-Valve ("Pop Valve"), or
	1 5-inch common Lever Valve.
	1 16-inch Glass Water Gauge.
	3 ½-inch Gauge Cocks.
	1 2½-inch Blow Cock.
	1 1¼-inch Feed Valve.
	1 1¼-inch Check Valve.
	1 1¼-inch Reverse Valve.
	1 5-inch Stop Valve.
	1 Bourdon Steam Gauge, 5-inch dial, iron body.
	All pipes, ells, tees, etc., required in fitting the boiler or use, and connecting it with the engine, heater and pump.
HEATER	One exhaust heater to be provided, and one erected in place, size No.—.
PLANING	The edges of all sheets must be planed and not chipped
FURNACE	To be 72 inches wide by 60 inches long, and to be built according to tracing.
GRATE BARS	1½ full set of bars and bearers to be furnished, bars to be in two lengths from door to bridge.
MASONRY	All masonry to be of the best material and done in a workmanlike manner, according to the tracing furnished.
TRACING	The tracing forms an integral part of this specification.
HYDRAULIC TEST	Before the boiler leaves the shop it must be subjected to a hydraulic pressure of 150 lbs. to the square inch without showing bad leaks, or evidences of undue strain, and be thoroughly examined and accepted by an Inspector of the American Steam Boiler Insurance Company of New York.

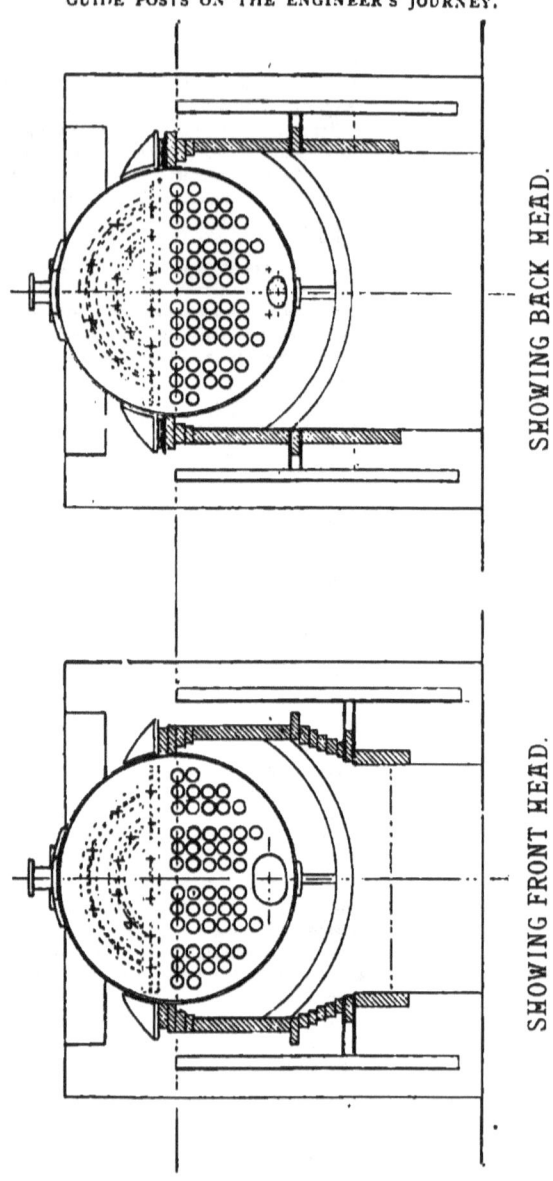

SUPERINTENDING THE CONSTRUCTION OF BOILERS.

An engineer who is superintending the construction of a boiler cannot be too careful. We will take it for granted that the owner of the works where the boiler is being built is thoroughly honest—so far all right—but he is liable to be deceived by others, so that it is not safe to take anything upon trust; he must depend upon his own careful inspection of material and workmanship, and he must take especial care that everything conforms closely to the specifications and plans in accordance with which the contract has been made.

In the first place, as soon as the plates arrive the thickness must be measured, and sample pieces should be tested for tensile strength and ductility. Should the test not come up to the requirements of the specifications, a new lot of iron must be obtained that will stand the test. While the plates are being sheared and punched a good opportunity is afforded for obtaining still further information in regard to the quality of the material. Should the blanks from the punch show dark marks on the edges, it is well to hold a few specimens edgewise on the anvil, and give them a smart blow with a hammer, as, if the iron is laminated, they will separate at the marks. Again, the distances from center to center of the rivet-holes should be carefully measured to see whether they are suitably proportioned to the thickness of sheet, and allow the right quantity of lap—too much lap being as great a fault as not enough; but if there is too much it must be reduced by the planer. Where there are flanges it should be seen that they are turned on the proper radius, and in a workmanlike manner, and the heads must present a plane surface, free from bulges or indentations.

It is not allowable to orm the tube holes in the heads in any other manner except by boring, and the sharp edges should afterwards be removed. The riveting should be sharply looked after, especially when done by hand; no rivets should be driven in half blind holes; the holes must be reamed and a larger rivet driven, and a number of these holes occurring in one seam on a single sheet is ample reason for rejecting that sheet. The caulking, if carelessly done, may weaken a boiler very materially from the first, and the broken skin presents an opportunity for the action ot corrosion, as well as assists in the production of grooving. Where nozzles and manhole frames are to be riveted, they should be first fitted closely to the shell, in order to prevent distortion and weakening of the shell. Reinforce plates should also be closely fitted, and well caulked after riveting. Especial care should be taken in regard to the bracing, that there is no slackness, or that no distortion is occasioned by them. The tubes should be carefully examined for splits and defects, and that they are properly placed and expanded, so that the head is not drawn out of shape by them.

The brackets should be examined to see that their bearing faces are in the same horizontal plane, and that they are properly placed. After the boiler is finished careful measurements from fixed points should be taken, and then blank joints should be made on all openings, a safety valve attached, and the hydraulic pressure put on, and new measurements taken to detect distortion. If the boiler has stood the pressure well, it should be emptied of its water, the manhole removed, and a thorough internal examination made of the braces. After this examination has been concluded, any bad rivet or seam leak that has been discovered during the water test should be made tight. This way of superintending the construction of a boiler takes time, and calls for a deal of care and patience, but it pays the owner of the boiler more than it costs, and in no other way can he be sure that he is getting what he pays for—a thoroughly reliable boiler.

As an owner would naturally insure his boiler, nothing should prevent him rom applying to the American Steam Boiler Insurance Co. for the plans and specifications of any boiler he may desire to have built, and to avail himself of the knowledge and skill of its trained corps of engineers and inspectors in superintending its construction.

BOILER SETTING.

There are various modes of setting boilers, and, in fact, every mason thinks that he knows all about the business; but there are certain points which should be regarded as of prime importance, ignorance or neglect of which may, and generally does, result in loss to the owner, through a consequent waste of fuel.

In the first place, the furnace must be roomy, in order to allow of the combustion of the principal part of the gases, before they are hurried over the bridge wall on their way to the flues. Then the walls of the furnace should be double, having a considerable air-space between them, to avoid loss of heat, and there should be a lining of fire-brick. The bridge wall should be concentric with the boiler bottom, and so arranged that the passage for the gases may have the proper ratio to the area of the grate, and it should be built hollow.

As gases are generated during the process of burning coal, suitable provision must be made to supply them with the air necessary for their combustion. Now this air can not be supplied through the grate bars, but it can be done in the urnace above the fuel, and also at and behind the bridge wall, the air required being supplied from the spaces between the walls of the furnace, and also from the hollow bridge. This air should be introduced in many fine streams, to promote its thorough admixture with the gases more rapidly, and its quantity depends upon the kind of fuel used. With a Return Tubular Boiler this extra provision of air is absolutely essential, as the tubes act as extinguishers, no flame being able to penetrate them to a greater depth than two or three inches, and hence any gas entering them unburned is wasted, so far as its evaporative power is concerned.

Of the many patented methods for effecting thorough combustion of the gases, those admitting air above the grates may be considered as decidedly the best, being founded upon sound chemical principles.

The space under the boiler, behind the bridge, and also the space at the back end of the boiler, must be considered as a prolonged combustion chamber, and it is during their passage through this chamber that the gases escaping unconsumed from the furnace must be burned, and to effect this, while they are at the proper temperature the air must be introduced near the bridge wall.

To insure a proper temperature, and thus promote their combustion, the boiler walls should be lined with fire-brick behind the bridge wall, and in the back connection also; as even when air is not introduced, a marked economy in fuel has been produced when a fire-brick lining throughout has been adopted.

And another thing is likewise of considerable importance, which is, to so arrange the masonry that the boiler may not be bound by it; as, when heated, the iron expands, exerting an almost incredible force, and, if room is not left for this expansion to take place in, cracks will be opened in the brick-work, and air will enter in quantities, impairing the draught, cooling the gases on their way to the tubes, and perhaps causing jets of flame to impinge so strongly upon the shell of the boiler as to repel the water and burn the plate, which may result in a rupture, and furthermore the masonry becomes shaky and insecure to such an extent that after a short time it must be entirely rebuilt.

It is sometimes as well, at the back end of the boiler, to choke off the upper part of the passages between the walls and the boiler shell, in order to distribute the gases more evenly among the tubes, and in many cases this has been found to have a beneficial effect.

In lining the wall of a furnace with fire-brick, a course of headers should be laid at a height, not exceeding 18 inches above the grate bars, for as the bricks, in contact with the burning fuel, deteriorate more or less rapidly, these headers will serve to sustain that part of the lining which lies above them while the others are renewed.

The bridge wall, on its furnace face, should be lined with alternate courses of headers and stretchers, for a thickness equal to the length of a brick, as it has to sustain very rough usage at times.

In regard to the position of the bridge, it may happen that a girth seam of the boiler is very near the place you have selected for it; now, in such a case, it is better to change the position of the bridge, so as to have the seam as nearly over the furnace edge of the bridge as possible, or better still, in the furnace itself, if it can be done without unduly lengthening the bars, otherwise the bridge must be so moved as to shorten the grates, and the area must be made up by increasing the furnace width. The greatest effect of the fire on the boiler bottom takes place at some point between the back of the bridge wall and a point about fifteen inches from it, and, as there is a double thickness of iron at a seam, there is great liability to burn the lap, should a seam lie at any point between the limits mentioned.

Now, in regard to the manner of supporting a boiler in its proper position in the masonry, opinions vary in different sections of this country. In the Eastern States, with single boilers, it is the general custom to secure cast-iron brackets to their sides, above the water line, which rest upon the walls, the rear brackets being supported upon rolls to allow of end motion in the boiler during contraction and expansion; but sometimes the furnace end of the boiler is supported upon the cast-iron front, while a pedestal supports the rear end. In the Western States a mud drum generally takes the place of a pedestal, with single boilers, the furnace end being supported by the front. When boilers are set in a battery, however, the general plan is to suspend them by rods from an overhead truss or arch, the extremities of which are supported by the walls, or by piers; and it is very usual to interpose a spring between the suspension-rod nut and the truss in order to allow for inequalities of strains.

Sometimes only the front ends of batteries are suspended from trusses, the rear ends being supported by the mud drums; but this is decidedly an objectionable plan.

Doors should be built in on the side or back walls of a boiler setting to permit of examination and cleaning, and for convenience in repairs, and suitable registers should be provided, communicating with the air-spaces in the double walls and the hollow bridge.

The passage to the chimney should be as direct as is possible, and supplied with a damper, but on no account should the boiler top be bricked over, either to form a flue, or to prevent loss by radiation, as both are highly detrimental, and the first is worse than useless, as has been decided long ago by the highest authorities.

Where the blow-pipe comes out from the bottom of a boiler it should be protected from the impact of flame and the heated gases by a half-round shield of fire-brick, or earthen sewer-pipe, and the horizontal portion, if practicable, should be laid in a recess formed in the bottom brick work of the combustion chamber.

It is well to prevent radiation from the top of a boiler by a covering of asbestos or other suitable material.

The bearer bars of a furnace should be so arranged that the grate bars may slope downwards from the furnace mouth towards the bridge, at the rate of one inch in one foot of length.

The height of the boiler bottom, above the grate, at the furnace mouth, should be 30 inches, and the rear end of the boiler should be 1½ inches lower than the front, in order to facilitate drainage.

The depth of the bottom of the ashpit, from the top of the grate bars at the furnace mouth, should be 30 inches, and the bottom of the ashpit should be well paved with good hard brick, laid on edge in sand and grouted. It should be level from the bridge to within one foot of the front, and should slope from thence upwards to the lower edge of the ashpit door, to facilitate the removal of ashes.

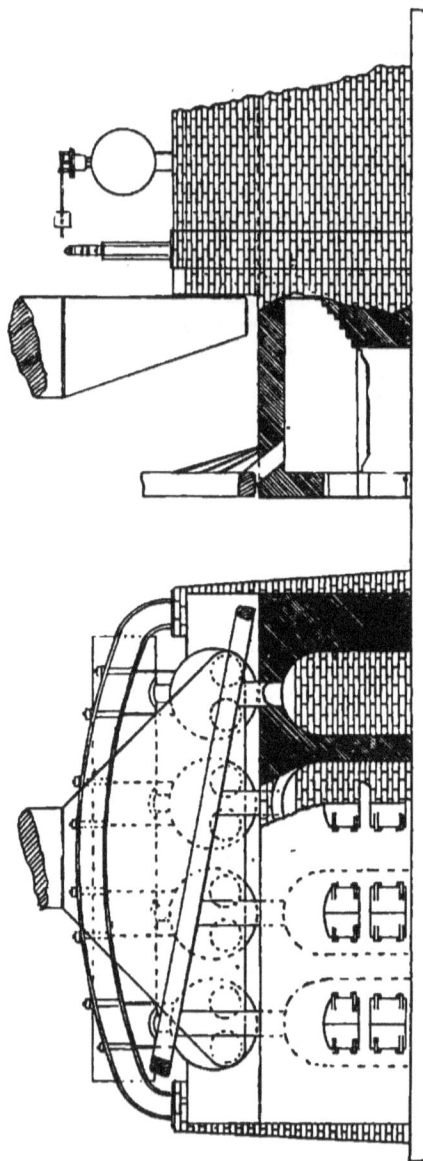

SHOWING EXTERNAL FURNACE OR OVEN.

GUIDE POSTS ON THE ENGINEER'S JOURNEY. 41

Wherever the brick-work comes into contact with the boiler, fire-clay or kaolin should be used nstead of ordinary mortar, and it should also be used with fire-brick; it must not be mixed to a stiff paste, as the joints should be as thin as possible.

Some plates, showing different methods of boiler setting, have been inserted in this work. Among others will be found one showing a boiler with a detached furnace, or oven. This style of furnace is used to some extent in the great saw-mill plants of Michigan, where the fuel is principally sawdust, slabs, and edgings. The plate explains itself, sawdust being continually fed, by spouts through the top of the ovens, on the fires of slabs and edgings below. This method claimed to be very efficient.

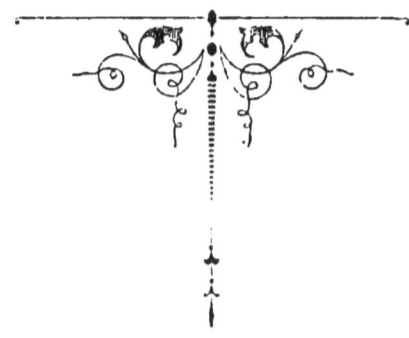

THE MANAGEMENT OF BOILERS.

It is of fully as great importance to manage a boiler properly as to have it properly designed and constructed. Poor management will soon render the best boiler useless.

From the time a boiler is set at work it is subject to destroying forces, which must be counteracted so far as is possible.

The fireman is the person under whose immediate care the boiler properly comes, and his duties, from not being generally understood, are apt to be undervalued, although they call for more knowledge than is generally supposed. To attend to his duty properly, he should know his business thoroughly, be prompt, reliable, careful, and last but not least, sober. There is a general adaptability for the situation about some men, but others might work a lifetime at the business and never learn it thoroughly. A certain handiness in the use of tools of various kinds is almost a necessity, although it is not necessary for him to have learned any trade, and he should have some pride in his work. It is not too much to say that a really good fireman is an almost invaluable man, and that he saves his wages to his employer more than twice over by the care and economy he exercises.

One thing is very important for owners to remember, which is this: *The fireman should have nothing to do except to attend to his boiler—Verbum sat.*

Before starting a fire under a boiler the fireman should see that there is plenty of water in the boiler, by trying the gauge-cocks; if there is not, he must at once fill the boiler to the proper level. Then he should see that the blow-cock is shut, that the hand-hole plates do not leak, or if they do he must make them tight, and he should look to see if there are leaks in the seams or at the tube-ends.

He should then see that the furnaces and flues are clear as well as the front connection, that the damper is open, that the ashpit is clear, and all ashes removed from in front of the boilers, and that the grate bars are in good order, for if not he must remove the badly warped or broken ones and substitute others.

He should then raise the safety-valve to allow air to escape while steam is forming, and should also open the upper gauge-cock to give notice when the steam begins to form, and he should also ease the stop valve, if there is one, to prevent its setting fast.

If the dampers and doors do not move freely, he should cause them to do so.

Everything being now ready, he should cover the grate bars with coal for about two-thirds of their length from the bridge, and should pile a little wood, cob-house fashion, on the open bars, and put a few lighted shavings, or oily waste, in the mouth of the furnace, partially close the furnace door, and wholly close the ashpit door. The coal on the bars prevents air from coming through them, and impairing the draught, while the partial opening of the furnace door supplies air to the burning wood, and directs the flame over the coal in the back end of the furnace, gradually heating it up to the point of ignition. After the wood is burning well, coal may be thrown upon it and the furnace door closed, the ashpit door being then opened. More coal is thrown on as soon as the fire will bear it, and the fire is gradually pushed back, till there is a full fire on the whole length of the bars.

Remember that the fire must not be hurried, except in case of an emergency; it must be "come up," as it is called, very gradually, and to do this in the best manner, *put on only a little coal at a time*, the fire will be more even for it. And remember, also, that if the coal is lumpy it must be broken, so that no piece is larger than a man's fist. In very long boilers, unless care is taken, steam may be making at the furnace end while the water in the other end of the boiler is hardly warm. There is a set of cylinder boilers, 72 feet in length, where the engineer gets up his fires so gradually that two or three days are required to raise steam on them to the working pressure, after they have been lying idle. As he is equally as careful in other matters pertaining to the plant under his charge, it may not be surprising that his engines and boilers, though old, are in the best possible condition. This plant is in Stamford, Conn. The fires should always be kept level and of a uniform thickness, with the exception that at the sides, corners, and at the bridge wall it must be enough thicker to prevent cold air from leaking through at these places With anthracite coal the thickness of the fire should be from 6 to 8 inches generally; with bituminous coal, from 8 to 10 inches; and with coke, from 10 to 12 inches.

The firing is only done properly when the fuel is consumed in the best possible way, that is, when no more is burned than is necessary to produce the amount of steam required, and to keep the pressure uniform; now to attain this end, complete combustion must be attained in the furnace, and this is going on when the fuel is burning with a bright flame evenly all over the grate. Blue flames, dark spots, and smoke are evidences of incomplete combustion, due to want of air, which ought to be supplied above the fuel in the furnace.

When the ashpit is uniformly bright the fire must be burning well; but if it is dark, in patches or wholly, there must be ashes and clinker on the grate bars, which should be removed at once. Ashes and clinker are also very apt to collect in the front corners of the furnaces, along the sides, and against the bridge wall, and these parts should be subject to special attention.

IMPROPER FIRING.

Experience only can teach a man how to fire, with the various kinds of fuel and under the different styles of boilers, in the best and most economical manner, though there are a few general principles applicable in all cases.

It is of primary importance that the boiler-room and its appurtenances, as well as the boilers and their fixtures, and the necessary tools, should always be kept clean. Every tool should have its place, and be kept there when not in use, and if broken should be repaired at once. The boilers should be fired at regular intervals, and lightly, and it has been found better in many cases, especially with wide furnaces, to fire one side of the furnace at a time; of this, however, each fireman must be his own judge, as no two boilers can be fired exactly alike and produce the same results.

When using a caking coal, a short time after firing it will be found necessary to make use of the slice-bar to break the cake into fragments, in order to get the full benefit of the coal, but care must be taken to keep the furnace door open no longer than is absolutely necessary; and firing also should be done as quickly as possible, to avoid contraction of the boiler bottom from the cool air entering the furnace. The fire should not be stirred any more than is necessary, in order to avoid the waste from small coal dropping through the bars. With a strong chimney draught it is as well to partially close the damper and also the ashpit doors when firing, to avoid contraction of the boiler sheets.

When the clinkers and dirt accumulate to an extent sufficient to clog the draught, the fire should be cleaned. Now, in boilers with wide furnaces it is better, perhaps, to clean only one-half at a time, and let the fire burn up well on that side before attempting to clean the other half. A fire, with a good quality of coal, should run about twelve hours without cleaning.

When there are several furnaces, all leading into the same chimney, they should be fired alternately, in order to keep the steam at a regular pressure and observe the greatest economy in fuel.

The practice of wetting coal before throwing it into the furnace can not be too severely condemned, as it is wasteful of heat and produces corrosion.

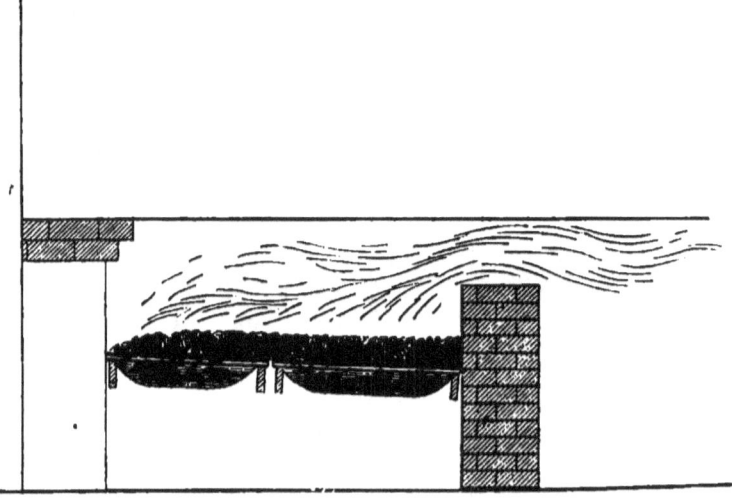

CORRECT FIRING.

The ashpits must be raked out frequently, and the air spaces between the bars must be kept free also. All ashes must be hauled to some distance from the boiler fronts and quenched, and then wheeled to the ash pile.

Never close the damper entirely while there is fire on the grates, as gas may collect in the flues, and an explosion might take place which would ruin the boiler. There is reason to believe that boiler explosions have been produced in this manner, —and or this reason alone it is considered better to draw the fires at night, instead of banking them, by many engineers. But it is the opinion of the author that a banked fire, properly kept up, is conducive to longevity of the boiler, because of the less amount of contraction and expansion induced, owing to difference in temperature. So far we have considered only the handling of fires; now let us give some attention to other matters of importance.

When a boiler is cold and filled with water, it will be found that, after the fire is lighted and steam is raised to the regular pressure, the gauge-cocks and water-glass show a higher water-level than before the fires were started, which is owing to the expansion of the water by heat. If now the throttle be opened and the engine started, the water will rise still higher in many boilers, showing a *false water-line*—for the water will drop to its proper level upon stopping the engine. This is owing to the violent ebullition going on in the boiler to supply the steam required, and which is being constantly drawn off—and it is more marked when the steam-room is small and the pressure high.

A boiler should have its feed water supplied regularly and continuously, and the water-line should be kept at a regular height, and there should never be less than three or four inches in depth over the highest part of the furnaces, flues, or connections exposed to the flames or hot gases, but it is very bad practice to carry the water too high in a boiler, and as a general thing the above-mentioned depth should not exceed five inches above the "fire" line, so-called.

In order to save fuel, a heater should be used, especially as it also increases the longevity of the boiler, and diminishes the tendency to leakage at the tube ends.

Blowing off steam at the safety-valve, or opening the furnace doors, to prevent a rise of steam pressure, causes loss of heat, which is synonymous with waste of fuel, and will never occur where a boiler is properly managed, except upon an emergency.

It is better to try the working of a safety-valve, which should be done once every day, by letting the steam rise gradually till the valve just "simmers," noting the pressure by the steam-gauge at that moment. Of course, you can raise it at any time by hand, but then you don't know *how much force you have exerted*, nor whether the steam would have raised it when it had attained the proper pressure.

The steam pressure should never be allowed to exceed its highest limit. If the steam-gauge shows that the steam pressure is rising rapidly, and that there is danger of exceeding the limit, water should be fed in at once, and the draught checked by the damper and ashpit doors; and if, in spite of this, the pressure should exceed the limit, the furnace doors should be opened, the feed put on more strongly, and the blow-cock partially opened, watching the water-line very sharply, as shown in the glass-gauge, and continually trying the gauge-cocks. But should it happen that the water is so low as to be dangerous, the pump must not be started, though if it is already in motion, it may be allowed to continue; the blow-cock must not be meddled with, but the furnace doors must be opened, and fresh coal, or better still, wetted small coal and ashes, must be thrown on the fire, and then in a few minutes the boiler will generally be cool enough to pump up. After such an incident, the tube ends should be he carefully examined in the upper rows for leaks.

The gauge-glass should be blown through several times every day, to see that it is not choked, and it is a very good plan to try the gauge-cocks every fifteen minutes during the day. Of course this necessitates a clock, but no fire room should be without one; it is as indispensable as the fire tools.

"Foaming" is a violent ebullition of the water in a boiler, which results in "priming," or the carrying of the water, in the state of fine spray, with the steam into the engine cylinder, often knocking out a cylinder head, and rapidly lowering the water level in the boiler, sometimes so much so as to be dangerous.

Foaming is generally caused by irregularity in firing or feeding; impure water, especially if it be greasy; contracted steam space; too small extent of area at the water-line; from the tubes being crowded together; the boiler not being clean ; the throttle or safety-valve being opened too suddenly; and, in marine boilers, changing the feed-water from salt to fresh, or the reverse. It is generally shown in the glass-gauge by a sudden rising or falling of the water, or by a boiling or showering of the water down through the glass, and by a peculiar sputtering sound given upon opening the gauge-cock. It can be overcome by partially closing the throttle, opening the furnace door, and feeding strongly; sometimes, however, it is necessary to blow out a little water from the boiler, but this should not be resorted to except in an extreme case.

In case the feed pump works badly, there may be several causes for it, such as:
Clogging of the strainer.
Broken or leaky valves.
Leaky packing.
Leaky or choked pipes.
The pump being hot.
The feed water being too hot.

The trouble should be ascertained as quickly as possible, and remedied at once.

A boiler needs cleaning out more or less frequently during each year, dependent upon the amount of impurities in the feed water.

When a fire is hauled, the stop-valve on the steam pipe should be shut; the ashes and cinders quenched and wheeled to the ash pile: the furnace and ashpit doors and the damper should be closed, and the steam blown off at the safety-valve till there is only about five pounds pressure, and then it should be lowered to its seat. On no account should the water be blown out of the boiler; it should remain there until it is cool, and then must be allowed to run out through the blow-cock. When the boiler is empty the man-holes and hand-holes may be opened, and the boiler must be examined internally, the scale must be knocked off with light blows of a pick, or scraped with bars and chisels, or loosened with angular wire chains, etc.; and the boiler must be washed clean with water. The scale should be removed as soon as possible after the water has been let out of the boiler, before it has time to dry and harden. Any repairs required by the braces, or at other points, should be made, and the boiler closed up. Then the blow-cock should be taken apart, the plug cleaned, oiled, and made to work water-tight and easily; the check-valves should be examined, and made tight; the feed-pipe and blow-out pipe should be examined and cleared of sediment, if any exists. The safety-valve should be examined, the pins taken out, cleaned and oiled, and the valve ground in, if it is leaky. The gauge-cocks should be examined, cleaned, and ground in, if necessary; the glass water-gauge should be examined and put into good condition; and advantage should be taken of this opportunity to send the steam-gauge to the maker and have it put into thorough condition.

The flues and connections should be swept, and the boiler bottom scraped with a wire brush; the furnaces should have the bars taken out, and the brick lining put into good condition, the ashpits cleaned and the bars replaced, renewing such as are badly warped or broken. The boiler is now ready for filling with water, which should be done at once, and then it should be examined carefully for leaks, which, if found, should be repaired at once, before the boiler is put into use again.

If, however, it is not intended to use the boiler for some time, it will be well to drain all the water out of it, and to dry it thoroughly by pans of charcoal, and then set a pan or two of lime into the boiler, and close it tightly to prevent oxidation.

A full set of fire tools, consisting of a shovel, slice-bar, T-bar, pricker, hoe, coal hammer and devil's claw, should always be kept in the fire-room, together with a broom and dust brush, and also a chipping hammer, a flat and diamond-point chisel, and wrenches to fit the nuts and bolts about the boiler, a monkey-wrench and screw-driver. A wheel-barrow is sometimes needed, also.

BOILER APPARATUS.

A safety-valve is designed to prevent the pressure in a boiler rom exceeding a certain limit, by opening when that limit is exceeded, and allowing the surplus steam to escape until the pressure has fallen a little below that limit, when it closes.

In order that it may work efficiently it is very necessary that it should be properly proportioned in all its parts.

Safety-valves are generally made with beveled seats, as shown in the accompanying figure. And unless a valve lifts clear of its seat, the opening it gives will be equal to the surface of the frustrum of a cone—having for the diameter of its upper base the diameter of the valve, $a\,b$, and for its slant-height the perpendicular distance between the lower edge of the valve and the seat : and for the diameter of its lower base the diameter of the seat, measured at the intersection of the perpendiculars, $b\,e$, $a\,f$, let fall from opposite points of the lower edge of the valve to its seat.

The bevel, or inclination, of the valve is the angle of inclination to a vertical line, $f\,c\,a$, $e\,d\,b$.

Now to find the amount of opening given by a valve with a beveled seat for any lift less than the depth of the seat, we have the following rule:

(1.) *Multiply the diameter of the valve by the lift, and that product by the sine of the angle of inclination, and that product by the number* 3.1416.

(2.) Multiply the square of the lift by the square of the sine of the angle of inclination, multiply this product by the cosine of the angle of inclination, and this last product by the number 3.1416.

(3.) Add the two final products, (1) and (2).

EXAMPLE.—The diameter of a safety-valve is 4 inches, the seat is ½ inch deep and has a bevel of 45°. What is the area of opening for a lift of $\frac{3}{16}$ inch ?

Diameter of valve....................................	4.0
Multiply by lift ($\frac{3}{16}$)................................	.1875
	.7500
Multiply by sine of angle of inclination (45°)............	.707
	.53025
Multiply by 3.1416....................................	3.1416
First product...	1.6658 +
Square of the lift.....................................	.035156 +
Multiply by the square of sine of angle of inclination (45°).......	.499849
	.01757 +
Multiply by cosine of angle of inclination (45°)...........	.707
	.0124 +
Multiply by 3.1416....................................	3.1416
Second product.......................................	0.039
Add first product.....................................	1.6658 +
Area of opening of valve in square inches..............	1.7048 +

DIAGRAM OF SAFETY VALVE WITH SEAT HAVING ANGLE OF 45°,

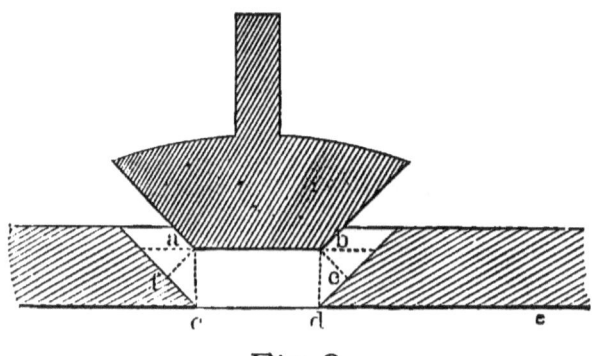

Fig. 2.

SHOWING AREAS FOR CALCULATION OF PRESSURES.

The most usual bevel for safety-valves is an angle of 45°, and the following is a short but correct rule given by Mr. R. H. Buel for a valve with its seat beveled to that angle:

(1.) *Multiply the diameter of the valve by the lift, and this product by the number 2.22.*
(2.) *Multiply the square of the lift by the number 1.11.*
(3.) *Add these two products.*

EXAMPLE.— What is the area of opening of a 2-inch valve, for ¼ inch of lift, depth of seat ⅜ inch, and bevel of valve 45°?

```
Diameter of valve...................................... 2.0
Multiply by the lift.................................... 0.25
                                                        .50
Multiply by 2.22....................................... 2.22
First product ......................................... 1.11

Square of the lift..................................... 0.0625
Multiply by 1.11 ...................................... 1.11
Second product........................................ 0.069375
Add first product..................................... 1.11

Area of opening of valve in square inches............. 1.179375
```

To facilitate calculations, a table of sines and cosines, from 20° to 50°, both inclusive, is annexed.

TABLE OF SINES AND COSINES.

Angle.	Sine.	Cosine.	Angle.	Sine.	Cosine.	Angle.	Sine.	Cosine.
20°	.342	.940	31°	.515	.857	42°	.669	.743
21°	.358	.934	32°	.530	.848	43°	.682	.731
22°	.375	.927	33°	.545	.839	44°	.695	.719
23°	.391	.921	34°	.559	.829	45°	.707	.707
24°	.407	.914	35°	.574	.819	46°	.719	.695
25°	.423	.906	36°	.588	.809	47°	.731	.682
26°	.438	.899	37°	.602	.799	48°	.743	.669
27°	.454	.891	38°	.616	.788	49°	.755	.656
28°	.469	.883	39°	.629	.777	50°	.766	.643
29°	.485	.875	40°	.643	.766			
30°	.500	.866	41°	.656	.755			

There are several rules for determining the proper area of a safety-valve for any boiler, and they are here given.

UNITED STATES STEAMBOAT INSPECTORS' RULES.

Allow one square inch of area of valve for every two square feet of area of grate, in the case of a common lever valve.

Allow one square inch of area of valve for every three square feet of area of grate, in spring loaded safety-valves adopted by the Board of U. S. Supervising Inspectors.

ENGLISH RULE.

For boilers with a natural draught, allow half a square inch of area in the valve for each square foot of grate surface.

PROFESSOR RANKINE'S RULE.

Allow a valve of $\frac{1}{1000}$ of an inch for each pound of water evaporated per hour.

RULE ADOPTED BY THE PHILADELPHIA DEPARTMENT OF STEAM-ENGINE AND BOILER INSPECTION.
(FRENCH RULE.)

1. Multiply the area of the grate in square feet by the number 22.5.
2. Add the number 8.62 to the pressure allowed per square inch.
3. Divide (1) by (2), and the quotient is the area of the valve in square inches.

The German government rule is based upon the steam pressure and heating surface, and requires a clear area of valve, or valves, after deducting for the wings or other obstacles, at the rate of so many *square lines* for each square foot of total heating surface, in accordance with the following table:

Working Pressure in Atmospheres	0 to 0.5	0.5 to 1	1 to 1.5	1.5 to 2	2 to 2.5	2.5 to 3	3 to 3.5	3.5 to 4	4 to 4.5	4.5 to 5
No. of sq. lines per sq. foot of Heating Surface	10	7	5.3	4.3	3.6	3.2	2.8	2.5	2.2	2.0

Note.—144 square lines = 1 square inch.

EXAMPLE.—A boiler has 900 square feet of heating surface, and is to work under a pressure of 75 lbs., or 5 atmospheres, the obstruction of the wings is 2 square inches. What is the required area? Clear area = 2.0×900 = 1,800 square lines; 1,800 ÷ 144 = 12.5 square inches; add for the obstruction of wings, and we have 12.5 + 2 = 14.5, area of the valve, which corresponds to a diameter of about $4\frac{3}{10}$ inches.

The English Board of Trade have prepared a table of safety-valve areas per square foot of grate area, varying with the different pressures, but they require that no valve shall be less than three inches in diameter, and they allow the area to be divided among two or more valves.

TABLE OF SAFETY-VALVE AREAS ALLOWED BY ENGLISH BOARD OF TRADE.

Boiler Pressure.	Area of valve per sq. foot of fire-grate.	Boiler Pressure.	Area of valve per sq. foot of fire-grate.	Boiler Pressure.	Area of valve per sq. foot of fire-grate.	Boiler Pressure.	Area of valve per sq. foot of fire-grate.	Boiler Pressure.	Area of valve per sq. foot of fire-grate.	Boiler Pressure.	Area of valve per sq. foot of fire-grate.	Boiler Pressure.	Area of valve per sq. foot of fire-grate.	Boiler Pressure.	Area of valve per sq. foot of fire-grate.	Boiler Pressure.	Area of valve per sq. foot of fire-grate.
lbs.	sq. in.	lbs.	sq. in.	lbs.	sq. in.	lbs.	sq. in.	lbs.	sq. in.	lbs.	sq. in.	lbs.	sq. in.	lbs.	sq. in.	lbs.	sq. in.
15	1.250	42	.657	69	.446	96	.337	123	.271	150	.227	177	.195				
16	1.209	43	.646	70	.441	97	.334	124	.269	151	.225	178	.194				
17	1.171	44	.635	71	.436	98	.331	125	.267	152	.224	179	.193				
18	1.136	45	.625	72	.431	99	.328	126	.265	153	.223	180	.192				
19	1.102	46	.614	73	.426	100	.326	127	.264	154	.221	181	.191				
20	1.071	47	.604	74	.421	101	.323	128	.262	155	.220	182	.190				
21	1.041	48	.595	75	.416	102	.320	129	.260	156	.219	183	.189				
22	1.013	49	.585	76	.412	103	.317	130	.258	157	.218	184	.188				
23	.986	50	.576	77	.407	104	.315	131	.256	158	.216	185	.187				
24	.961	51	.568	78	.403	105	.312	132	.255	159	.215	186	.186				
25	.937	52	.559	79	.398	106	.309	133	.253	160	.214	187	.185				
26	.914	53	.551	80	.394	107	.307	134	.251	161	.213	188	.184				
27	.892	54	.543	81	.390	108	.304	135	.250	162	.211	189	.183				
28	.872	55	.535	82	.386	109	.302	136	.248	163	.210	190	.182				
29	.852	56	.528	83	.382	110	.300	137	.246	164	.209	191	.181				
30	.833	57	.520	84	.378	111	.297	138	.245	165	.208	192	.181				
31	.815	58	.513	85	.375	112	.295	139	.243	166	.207	193	.180				
32	.797	59	.506	86	.371	113	.292	140	.241	167	.206	194	.179				
33	.781	60	.500	87	.367	114	.290	141	.240	168	.204	195	.178				
34	.765	61	.493	88	.364	115	.288	142	.238	169	.203	196	.177				
35	.750	62	.487	89	.360	116	.286	143	.237	170	.202	197	.176				
36	.735	63	.480	90	.357	117	.284	144	.235	171	.201	198	.176				
37	.721	64	.474	91	.353	118	.281	145	.234	172	.200	199	.175				
38	.707	65	.468	92	.350	119	.279	146	.232	173	.199	200	.174				
39	.694	66	.462	93	.347	120	.277	147	.231	174	.198						
40	.681	67	.457	94	.344	121	.275	148	.230	175	.197						
41	.669	68	.451	95	.340	122	.273	149	.228	176	.196						

Safety-valves are of three kinds, known as common lever, deadweight, and spring-loaded. The common lever valve is in more general use than the others, but it is imperfect in its action, as it will not seat until the steam pressure has fallen several pounds below the point at which it is set to open; and, again, owing to the varying angle of the lever in a sea-way, the load on the valve is not constant, thus wasting much steam, so that its use is practically confined to stationary boilers. The plate shows the valve as approved by the Board of U. S. Supervising Inspectors of Steam Vessels.

The principle of this valve is that of a lever of the third order, the power to raise it being applied between the fulcrum and the weight. If the lever itself weighed nothing it would be a very simple matter to calculate the position of the weight, but as both the lever and valve are material, their weight forms a very essential part of the calculation. The manner of making this calculation can be best explained by giving an example, illustrated by a diagram, as follows:

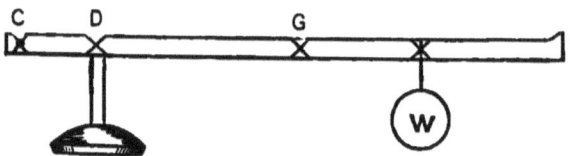

Let AB represent a lever of a safety-valve 40 inches long, and weighing 10 pounds. Let the distance from the fulcrum, C, to the point of application, D, of the power of the steam tending to raise the valve, V, be 5 inches; the distance from C to G, the center of gravity of the lever, be 20 inches. Let the diameter of the valve be 4 inches, and the weight be 80 pounds, where shall the weight be placed on the lever, *i.e.*, at what distance from the fulcrum, in order that 50 pounds pressure per square inch may be kept upon the valve, supposing the valve itself to weigh 5 pounds?

As the valve is 4 inches in diameter, its area is 12.56 square inches, and multiplying this area by 50 lbs., the pressure, we have a total pressure or power of 628.32 lbs. on the valve, which we must keep in place by the weights of the valve and of the lever, and by the weight on the lever. The valve acts as a dead weight, so by subtracting its weight, 5 lbs., from 628.32 lbs., we find we have only 623.32 lbs. resistance to be overcome, but this resistance is applied at 5 inches from the fulcrum, so we multiply by 5, and we have 3,116.6 lbs. The lever weighs 10 lbs., and acts with a leverage of 20 inches, thus making 200 lbs. of weight tending to keep the valve closed, and this subtracted from 3,116.6 lbs. leaves 2,916.6 lbs. to be neutralized by the weight. Now if we divide 2,916.6 lbs. by 80 lbs., we shall have 36 47, which is the distance in inches at which the weight must be placed from the fulcrum.

To test a lever safety-valve, when there is no steam upon the boiler, it is necessary to know where the center of gravity of the lever is situated. This may be found by disconnecting the lever and trying it upon a knife edge, at right angles to its length, until the position is found where it balances, which must be marked, and its distance from the fulcrum measured; then the lever must be weighed, as also the valve, and the weight used as a load; and the diameter of the valve must be measured also. From these data, as shown above, all the necessary calculations can be made. But there is a practical method of ascertaining the weight on a safety-valve, which is as ollows:

Secure the valve-stem of the safety-valve to the lever with a wire, but not rigidly; then affix a loop, into which you pass the hook ot an accurate scale beam; then secure the scale beam so that it takes the weight of the lever and valve when weights are applied to it, and weigh the apparatus. The weight, as given on the scale, divided by the area of the valve in square inches, will give the pressure in pounds at which the steam will raise the valve. For small valves an accurate spring balance may be used.

When a boiler is under steam, an accurate guage can be attached to it, and the valve weight adjusted by means of that, and, in fact, it is the only accurate method.

A deadweight valve is a simple valve with a long stem, and is weighted by iron disks having a hole at their center for the valve-stem to pass through, in order to secure them in place. In a sea-way they have the same defects as the common lever valve.

They can be tested by weighing the valve and the disks, and then dividing the total weight by the valve area, or, still better, by the use, as before mentioned, of an accurate steam-gauge.

In a spring-loaded valve the valve is kept on its seat by the pressure of a spring. This spring, no matter in what position it may be placed, always blows off at the pressure for which it is set. There are many patents for different styles of this valve, and the best of them, approved and recommended by the Board of U. S. Supervising Inspectors of Steam Vessels, give so much larger a discharge opening than a common-lever valve of equal diameter, that it has been granted 50 per centum of allowance of efficiency in excess of that of the lever-valve,—that is, one square inch ol area of valve only is required for every three square feet of grate. These valves have a valve seat beveled to 45°, and are nickel plated, or have nickel seats, thus avoiding corrosion or "sticking."

A great advantage possessed by spring-loaded valves lies in the fact that firemen can not overweight them by hanging old pieces of grate bars, etc., on them, nor shove them down on their seats, which can be, and often is, done where a common-lever valve is used; and, again, they will close before the pressure has dropped a pound below the point at which they are set, thus avoiding waste of steam. They must be set by the use of a steam-gauge.

Safety-valves should be kept clean, and should be frequently tested to see that they work freely and are correctly weighted.

STEAM GAUGES.

It is absolutely necessary that every boiler should be provided with a trustworthy steam-gauge for indicating the pressure at any and all times, and this gauge should be connected directly to the boiler, and not to the steam pipe, in order to avoid fluctuations in pressure.

The kind almost universally used at the present day is the dial gauge, the mercurial gauge not being suitable for the high pressures now carried upon boilers. The principle of action of the dial gauge is the tendency of a flattened curved tube, closed at one end, to become straight when subjected to internal pressure, and when well made it is very reliable. It was invented by M. Bourdon, but the patent has lapsed, and many poorly constructed and unreliable articles have been since sold as Bourdon gauges.

In all metallic spring gauges the connecting pipe should be fitted with a "syphon," ust below the gauge, filled with water, which serves to transmit the pressure and prevents the steam rom coming in contact with the spring. In cold weather care must be taken not to let this water freeze.

Suitable cocks must be fitted to the connection pipes, n order that the steam may be shut oft and the gauge examined rom time to time.

Where there are several boilers in a battery, each boiler should be provided with a separate steam-gauge, which must be connected to the boiler directly.

One great fault with this gauge lies in the act that there is generally so much ornamentation and lettering on the dial that the figures on the pressure scale are frequently almost illegible.

WATER GAUGES.

It is fully of as great importance to know the height of the water in a boiler as to know the amount of steam pressure, and to ascertain where the water level in a boiler is at any time, we make use of gauge-cocks and glass water-gauges.

Gauge-cocks are more generally used, and no boiler should be without them. There should be three gauge-cocks on each and every boiler, one at the lowest point to which the water can fall without great danger, and this is usually, in a Return Tubular boiler, at a height of one (1) inch above the highest row of tubes, another at about six (6) inches above this, and one at a distance intermediate between them. These should be tried frequently, and always kept clear and in good working order.

Glass water-gauges were introduced to obviate the trouble of trying the gauge-cocks when wishing to know the height of the water in the boiler. A gauge of this kind consists of a glass tube, the top and bottom of which, by means of suitable fittings, communicate with the steam and water spaces of the boiler. The level of the water in the glass is taken to be the same as that of the water in the boiler, and it is always before the eyes of the engineer or fireman. To provide for renewing, cleaning, or repacking, cocks should be arranged to shut off communication with the boiler, and there should be, also, another cock, at the bottom of the gauge, for emptying the glass when it is desired to drain off the water or to ascertain whether the gauge is working properly. Provision ought also to be made for cleaning out the steam and water passages while the boiler is at work, in the event of their becoming choked. In the selection of a gauge, or of gauge-cocks, take care that the passages are not too small; nothing less than one-half inch is admissible, and with dirty water, or where much scale is produced, five-eighths is a preferable diameter of bore.

BLOW-OUT APPARATUS.

The bottom blow-out apparatus should be used simply for emptying the boiler, and it should be so fitted as to do this completely. It is of the utmost importance that its tightness can be depended upon, as a small amount of leakage, continued for a length of time, might be the cause of "low water," with consequent overheating of the plates and destruction of the boiler.

A plug cock is the simplest, surest, and at the same time most durable valve that can be fitted to a blow-out pipe. If it leaks, the leak is manifest, and the position of the handle always shows whether it is open or closed.

In addition to the bottom blow-out apparatus, every boiler should be provided with means for blowing out water from the surface, in order to remove the fine particles of foreign matter floating there, which afterwards settle and consolidate as scale on the heating surfaces.

It consists, in its simplest form, of a pan, or conical scoop, near the surface of the water, but below it, connected with a pipe passing through the boiler shell, on which is a cock, or valve, for regulating the escape of the water laden with the impurities deposited in the pan. There are patented apparatus for this purpose which are well designed and easily fitted to a boiler.

FUSIBLE PLUGS.

In some States the insertion of fusible plugs at the highest fire-line in boilers is compelled by law, under a heavy penalty. The plugs, however, need frequent examinations, and to keep them in efficient condition their surfaces must be often scraped bright, and in spite of the utmost care they are not reliable.

INJECTORS.

An injector is always a necessity where no heater is provided for the feed-water, as it prevents the contraction of tubes and plates where the feed-water impinges upon them.

PUMPS.

A steam pump is always necessary, even when an injector is fitted, and one with a hand motion is preferable. It should not be so large as to render it difficult to feed the boilers continuously at a slow rate of speed.

HEATERS.

These are a most reliable adjunct to a boiler. In addition to the economy in fuel arising from their use, owing to the high temperature imparted to the feed-water (every increase of temperature in the feed-water of 11° corresponds to an economy in fuel of 1 per centum), there is a still greater benefit obtained in the most improved varieties, viz.: the removal of the greater part of the impurities from the water before it enters the boiler, thereby preventing the formation of heavy incrustation or deposit. And, again, some of these heaters are so arranged that they act as condensers also, which adds still more to the economy to be attained by their use.

STEAM DOMES.

These are frequently attached to boilers under the mistaken idea of increasing the steam room and procuring dry steam; but unless tubes conveying air or gases at a high temperature are led through them, they effect nothing in the way of drying steam, and the extra amount of steam room they afford is of very little value; but they are a positive detriment to any boiler, from the extra amount of surface they afford for radiation. It has been estimated that a square foot of uncovered boiler-top surface will condense steam enough every hour to develop one-third of a horse-power, and the additional surface afforded by an ordinary dome of 30 inches by 30 inches, is sufficient to condense steam enough to have developed more than eight horse-power.

A dry-pipe, so-called, perforated on its upper side, and running nearly the whole length of a boiler, is much preferable.

REVERSE VALVES.

These valves are made to open inwards, towards the interior of the boiler, by atmospheric pressure, and are directly opposite in action to a safety-valve. They are a useful adjunct, and are easily fitted on the top of the water column, or on the shell of the boiler itself.

MUD DRUMS.

These are considered as almost indispensable west of the Alleghanies, but they are a fruitful source of trouble. They are made of varying sizes, from 16 inches up to 30 inches in diameter, and are usually placed athwartships, under the rear end of a battery of boilers, with a short, upright connection to each boiler of considerable diameter. The drum is supported on iron pedestals rising from a solid foundation, and they are often arched over with brick to prevent direct contact of the flame and heat. The object of having a drum is threefold: to equalize the water level in the boilers, to serve as a receptacle of mud and dirt, and to distribute the feed evenly, as well as to warm it, as the feed-pipe is generally led into one end of the drum at its upper side. Upon examination the upper part of the drum will be found much warmer than the bottom, and the result, as might be expected, is nequal contraction and expansion, producing leaks in the ring-seams and neck-flanges, as well as corrosion at those places, together with grooving of the boiler, and internal corrosion of the sheets of the drum. It has been found that the larger is the diameter of the drum, the worse are these defects.

By using a small, seamless mud drum, and not feeding through it, these defects very rarely occur.

GRATE BARS.

When a plain grate bar is used, it is better to cast three or our together in a block, so that it will require about three of them in the length of a long furnace; as when made in this way they are much lighter and less liable to warp and burn out, and give a much better distribution of air to the fuel. There are many patent grates, rocking and otherwise arranged, which are very efficient and some of them are of the greatest assistance in cleaning fires.

DAMPERS.

These are of great use, but they should be so fitted that, when closed, there will be a space around their outer edge to allow of the escape of gas to a moderate extent. This is a necessary precaution where the habit of banking fires is prevalent.

DEFECTS.

Some of the more prominent defects of boilers are here mentioned, together with their causes and the usual remedies therefor:

LOOSE RIVETS.—This is generally the effect of overheating. They should be cut out and new ones driven; but in case the rivets are too small for the thickness of the plates, and especially in a girth seam, the rivets must be replaced by others of a larger diameter.

BLISTERS.—These are due to imperfect welding in the manufacture of the plates. They should be trimmed off to ascertain their extent and thickness, and, if of small area and slight thickness, are not dangerous; but if the contrary is the case, or if the plate is cracked under the blister, they must be cut out, and a "hard patch" put on inside the boiler to avoid making a pocket for the collection of sediment. Sometimes this patch will be found to leak, and caulking will not stop it, in which case, on removing the patch previous to renewal, it will generally be found that there are cracks under the heads of the rivets.

BURNT PLATES.—These are due to "low water," to a deposit of sediment or scale, to continued impact of flame caused by leaks of air through the masonry, and when a seam is just back of a bridge wall; but sometimes it is caused from an incrustation, or soap formed from oily matter. The place should be cut out and a hard patch put on.

BAGGING, BUCKLING, or BULGING of plates, sometimes orming a pocket, generally occurs from the overheating of the iron in consequence of deposits of oil, sediment, or scale. Sometimes it occurs when the boilers are clean, and then it is the result of impact of flame, or it may be caused by the unequal expansion of the various laminæ of the plates under their daily usage in case there are no signs of overheating. It is very liable to occur in the flat fire-sheets of the sides of fire-box boilers when the stays are spaced too far apart. The usual remedy in this case is to put in an additional stay-bolt between each four stays already there. The presence of a bulge on the bottom fire-sheet of a boiler is not necessarily dangerous, but it must be watched carefully, and its surfaces kept clean, and, at the first indication of weakness, it must be heated and forced back into place, or else cut out and a hard patch put on.

SEAM LEAKS are generally considered fair evidence of overheating, especially in the lower half of the girth seams. They should be carefully examined, the rivets having been cut out, and if any evidences of cracks under the heads of the rivets are found, the seam should be cut out also and a patch put on. Sometimes a leak occurs in a seam from the rivets being too small, or from the lap being too great. In the former case the rivets should be replaced by larger ones, and in the latter the seam must be chipped and caulked.

CRACKS in plates may be due to overheating, or to unequal contraction and expansion, or to letting cold water strike on a hot plate, and this last often happens in hurrying to clean a boiler.

As a general rule, it is safe to say that accidents from the overheating of boiler surfaces do not occur at the moment of overheating, but at some subsequent period. How soon depends upon the extent of the overheating. Steam boilers can be used with almost perfect security if proper attention is paid to them, requiring merely a careful observance of natural laws, and the constant exertion of that much-neglected faculty called common sense.

Neglect of the masonry in the setting of a boiler is often the cause of external corrosion, and cracks or loose bricks should never be allowed. Nothing but fire-clay or kaolin should be used to cement the bricks wherever they come in contact with the boiler.

The braces of a boiler need careful attention. All scale and rust should be removed from them, and they should be sounded with smart blows of the hammer at the extremities to detect slackness or breaks. A broken brace should always be repaired as soon as discovered. The gauge-cocks, feed and blow-cocks, when found to leak, should be ground in as shortly thereafter as possible; and, if they are choked, the obstructions must be removed. It is well to open the blow-cock a little once every day to prevent its setting fast. If a back stave is broken, a new one must be put in its place at once, or the walls may begin to bulge or crack.

The cast-iron front often cracks, and it should be repaired by bolting to it a piece of iron plate or bar to prevent widening of the crack or sagging, unless a new section can be readily procured and substituted for it, which is preferable.

Leaky feed and suction pipes must be repaired by parceling and serving, if of copper; but if the pipe is of iron, it should be thrown aside and a new length substituted.

If a gauge-glass breaks, shut off the water first, and the steam afterwards, to avoid being scalded

If a bolt blows out or a tube splits, drive in a pine plug till you have time to repair it properly, which should be after work is over for that day.

The brick lining of the furnace is always getting out of order, especially at the front and at the bridges; don't wait too long before you make the necessary repairs.

Remember that prompt repairs save long bills.

Some time ago a pair of new double-flue boilers, connected by a mud drum, were erected in a mill in Southern Ohio. After a very short time it was found that the girth seams of the drum were leaking badly on the upper side. The necessary repairs were made at once, but very soon afterwards bulgings and cracks appeared in some of the sheets along the top, and cracks showed in the top edges of the girth seams, and the leakage was worse than before. A very close examination was then made of the drum, and it was found that the ring courses composing it were so arranged that the necks forming the connections with the boilers were placed on the "small courses," as shown in the accompanying engraving, leaving a small chamber at each of the "large courses" to fill up with steam, thus preventing the contact of water, and allowing the heat to have a destructive effect upon the iron. Another drum was substituted, having the necks on the large courses, since which time no further trouble has been experienced. So much for carelessness in construction.

EXPLOSIONS.

Explosions of steam boilers are due to defective material, natural deterioration, defects of construction, or to improper management; they take place when the resistance is less at some point than the pressure to which it is subjected, and may happen even when the pressure is very low.

The steam pressure in a boiler is never increased otherwise than by the natural cumulative effect of the furnace. There exists no phenomenon that has power to suddenly increase the power in a boiler. The action which has been attributed in this respect to the "spheroidal state" and to the "superheating of the water" has not been confirmed by observation.

The explosion of a boiler is not an instantaneous action, though seemingly so; it is a well-defined and rapidly succeeding series of operations. The rupture commences at the point where the resistance offered by the material is less than the strain to which it is subjected, and it extends into the adjoining parts when these parts are too weak to sustain this increased strain that the rupture already made brings to bear on them, together with the shock due to the motion that the edges of the fracture make while seeking a new state of equilibrium.

The number and direction of the ruptures depend especially upon the resistance of the parts adjacent to the first fracture. A rupture, even when of considerable extent, does not produce an explosion if the adjoining parts possess sufficient strength.

In case of an explosion the steam pressure does not fall the instant that the rupture takes place. On the contrary, the pressure continues very nearly constant up to the time when all the water has escaped from the boiler.

An explosion is so much the more terrible as there are more fractures made prior to the moment when the boiler is entirely emptied of its water.

It is very dangerous to let the water get so low in a boiler that the plates become red-hot, because the softened plate will tear open, and may produce an explosion if the red-hot part is of large extent, or if the adjoining parts do not offer sufficient resistance.

When water is fed into a boiler where the water is too low, it almost invariably lowers the pressure of the steam. So that it is always dangerous to introduce feed before having dampened the fires with wetted small coal and ashes, or having drawn them, because the water injected quiets the ebullition, and increases the surface exposed to the heat.

There is not, nor can there be, any connection between the explosion of steam boilers and the phenomenon known as "the spheroidal state."

Even when "low water" does not cause the plates to be heated to redness, it causes leaks in the seams and at the tube ends, and fractures also by reason of the unequal contraction and expansion produced.

It is dangerous to empty a boiler when the flues or tubes are still hot, or to fill a boiler with cold water before it becomes sufficiently cool, or to wash it out while it is still warm, for such action causes fractures of the transverse riveting in such a manner as may not always be shown by leakage, and this defect may very easily produce an explosion when next the fires are lighted, or in a short time afterward.

It is also dangerous to fire up under a boiler too rapidly, as when the draft and combustion are sufficient for a "white heat," the plates, no matter how good they are, can not resist with certainty.

Explosions are generally due to unseen defects, because no one has tried to discover them.

In Newark, N. J., a short time ago, the water got low in a new and strong boiler from inattention on the part of the engineer, and many of the tubes required resetting, but no inspection of the boiler was made or attempted. A few months afterwards it exploded with violence shortly after the feed had been put on, causing serious damage.

Overheating of a boiler may also be caused by accumulation of scale or sediment, or other foreign matter, on the furnace plates, flues, or tubes and heads, or by the metal being too thick near the fire, and by defective circulation.

Defective circulation is generally due to the design of the boiler, from its having too cramped water spaces; from the connection being impeded by the overcrowding of tubes, or placing them too close over the furnace crowns; and from having too large a body of dead water lying below the heating surfaces, which causes unequal contraction and expansion. In one case that came under the writer's notice, a large boiler on an ocean steamship was fired up the day previous to her regular sailing day, and steam was raised to ten pounds pressure; but the water bottoms were sufficiently cool for a person to keep his hand on them, and a few hours later one of the girth seams opened out for about three or four feet in length. This was due to the imperfect circulation in the boiler, due to the large body of dead water lying under the heating surfaces, causing unequal expansion and contraction.

Fairbairn states that the efficiency and safety of a boiler depend as much upon the efficacy of the circulation as they do upon the strength and disposal of the material of which it is composed.

The hydraulic pressure applied to a boiler does not show us whether it is dangerous to use it or not. Many boilers have exploded when corrosion has reduced the metal so much that a smart blow from a ball-faced hammer would have knocked a hole in the sheet, while the boiler has given no evidence of weakness under the hydraulic pressure.

Another source of explosions is due to the sudden and unequal expansion of the metal of large cast-iron stop-valve chambers when steam is suddenly let on in cold, frosty weather through opening the valve, water of condensation being present. Such explosions have taken place with only ten pounds pressure.

The writer is acquainted with two explosions produced on steamboats, where there were two boilers connected by stop-valves, which were shut. In each case one of the boilers had just been repaired, while steam had been kept upon the other; as soon as the repaired boiler began to make steam the communication with the other was opened wide, and an explosion, resulting in loss of life, took place, owing to the sudden shock and reaction.

Boilers can also explode from a preliminary explosion of gas in the furnace or flues.

A thorough internal and external inspection of boilers by a person skilled in the profession is the only means of ascertaining their condition. Each and every part of a boiler may contain dangerous defects, and an examination is only finished when every part of the boiler has been carefully inspected.

When a plate is covered with soot or incrustation most of the defects can not be seen; therefore it is highly important that boilers should be kept as clean as possible, as well externally as internally.

It is a bad plan to hasten the cooling of a boiler, especially if it is a long one, and it is very injurious, also, to let the water out of a boiler until the tubes or flues are sufficiently cooled. Experience has shown that out of every ten accidents or explosions that happen, only one occurs to a regularly inspected boiler. From what has been said it is plain that a boiler may burst and not explode, but that an explosion is always preceded by a bursting to which the explosion is a consequent.

A BAD RUPTURE.

The various steps in an explosion, which have been before mentioned, may be defined as follows:

FIRST.—A fracture in a plate followed by a rending.

SECOND.—A violent outburst of water and steam.

THIRD.—A fall in pressure.

FOURTH.—Portions of water are propelled with great violence against the shell of the boiler, and shattering it, by the expansive force of the steam disseminated throughout the body of the water.

FIFTH.—The steam generated from the liberated water imparts a high velocity to the fragments, converting them into projectiles, and thus spreading ruin and destruction around; and also widely scattering the particles of water not converted into steam.

It is generally said, by those not thoroughly conversant with steam, that, in case of an explosion, the water must have been low; and they point, in corroboration of their statement, to the fact that no water is found scattered round. Now nothing is more certain than that all of the water contained in the boiler can not be converted into steam when an explosion occurs, for 965 units of heat are required to boil water into steam from and at 212° Fahrenheit; while at 140 lbs. pressure the temperature would be 361°, or 149 units higher than 212°. Now, 965 divided by 149 is 6½ nearly. That is to say, of every 6½ lbs. of water in the boiler *one pound* only will be converted into steam of atmospheric pressure, while 5½ lbs. of *water* will be scattered in the air, mixed with the escaping steam.

In the village of Cortland, New York, a steam-boiler exploded; there was plenty of water in it at the time. Persons alarmed by the noise, and hastening to the scene of the disaster, were wet by the water, falling like rain, at a distance of half a mile off.

From various experiments and investigations the following conclusions have been arrived at:

1. A violent explosion may take place in a boiler when there is plenty of water in it.
2. A moderate pressure of steam may produce a terrific explosion when there is plenty of water.
3. That a boiler may explode under steam at a less pressure than it has stood without apparent injury from a water pressure.
4. A rupture will be followed by relief of pressure, with or without explosion, as the fracture is extended or otherwise.
5. That an explosion rarely occurs in an externally fired boiler from "low water."

In examining into the cause of a boiler explosion it is generally necessary to determine the initial point of rupture, which often will solve the whole question; but before making a final decision we should be prepared to show:

1. That the cause can exist in the case in question.
2. That it is competent to produce the results ascribed to it.
3. That no other known cause can produce these effects.

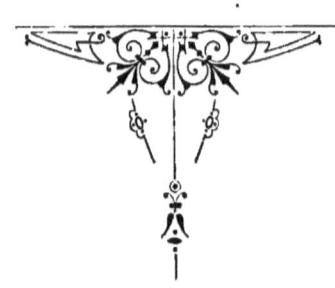

FUEL.

There is often required, for the use of the boilers of the average manufactory, nearly twice as much fuel as would be necessary if the plant were of a better type and more skillfully managed; and in some places the waste is greater than this.

The way in which this waste takes place may be stated generally as follows:
1. The engines require too much steam to develop the power required.
2. The boilers are badly designed, improperly set, or too small.
3. The coal used is of poor quality, or improperly housed.
4. The firemen are careless or ignorant of their business.

The proper course for an owner to pursue, in case he suspects waste, is to employ a skilled expert engineer to examine his plant thoroughly and report upon the defects discovered, stating the remedies therefor; and then to act upon that report as he would in any other matter of business.

Experiments have shown that coal loses from *ten* to *forty* per centum of its evaporative effect from being exposed to the weather. It should always be kept under cover, and the building should be of brick and closed in.

It is generally conceded that $2\frac{1}{4}$ lbs. of good dry wood are equivalent in evaporative effect to one pound of good coal; but it must be remembered that wood requires a roomier furnace than coal, and also that the spaces between the grate bars must be larger.

The fuel value of the same *weight* of different woods is very nearly the same—that is, a pound of hickory is worth no more for fuel than a pound of pine, assuming both to be dry.

If the value be measured by weight, it is important that the wood be dry, as each ten per centum of water or moisture in the wood will detract about twelve per centum from its value as a fuel.

The weights of one cord of different woods (air dried), as well as the fuel value in comparison with coal, is as follows:

1 Cord Hickory or Hard Maple weighs 4,500 lbs. and is equivalent to 2,000 lbs. of Coal.
1 Cord White Oak weighs 3,850 lbs. and is equivalent to 1,715 lbs. of Coal.
1 Cord Beech, Red Oak, Black Oak weighs 3,250 lbs. and is equivalent to 1,450 lbs. of Coal.
1 Cord Poplar (white wood), Chestnut, Elm weighs 2,350 lbs. and is equivalent to 1,050 lbs. of Coal.
1 Cord the average of Pine weighs 2,000 lbs. and is equivalent to 925 lbs. of Coal.

The following comparative results of the values of anthracite, semi-bituminous, and bituminous coals may prove of interest:

TABLE SHOWING THE VALUE AND PROPERTIES OF VARIOUS KINDS OF COAL.

DESIGNATION OF COAL.		Weight per cubic foot by experiment.	Cubic feet of space required to stow 1 ton.	Volatile combustible matter in 100 parts.	Fixed carbon in 100 parts.	Earthy matter in 100 parts.	Pounds of steam to 1 of coal from 212°.	Total waste (ashes and clinker) from 100 of coal.	Weight of clinker alone from 100 of coal.	Av. weight in pounds of unburnt coke left on grate after such experiment.	Steam from 212° from 1 of combustible matter.
Beaver Meadow, slope No. 3	Pa.	54.93	40.78	2.38	88.94	7.11	9.21	11.96	1.01	112.4	10,462
Beaver Meadow, slope No. 5	Pa.	56.19	39.86	2.66	91.47	5.15	9.88	6.74	0.60	61.2	10,592
Forest Improvement	Pa.	53.66	41.75	3.07	90.75	4.44	10.11	6.97	0.81	40.2	10,807
Peach Mountain	Pa.	53.79	41.64	2.96	89.02	6.13	10.06	6.97	1.08	26.6	10,896
Lehigh	Pa.	53.32	40.50	5.28	89.15	5.56	8.93	7.22	3.03	36.1	9,626
Lackawanna	Pa.	48.89	45.82	3.91	87.74	6.35	9.79	8.93	1.24	57.2	10,764
Lyken's Valley	Pa.	48.56	46.13	6.88	83.84	9.25	9.46	12.24	4.40	18.0	10,788
Beaver Meadow (Navy Yard)	Va.	55.08	40.65			8.10	9.08	8.10	1.40	107.1	9,881
Natural Coke of Virginia	Va.	46.64	48.03			11.83	8.47	18.46	5.31	60.9	10,389
Coke of Midlothian Coal	Va.	32.70	68.50	12.44	75.08	16.55	8.63	16.54	10.51	53.2	10,343
Coke of Neff's (Cumberland) Coal	Md.	31.57	70.95			13.34	9.00	13.34	3.55	43.7	10,381
Mixture 1-5 Cumberland and 4-5 Meadow		54.29	41.26			8.88	8.86	8.88	4.91	9.5	9,775
N. Y. and Maryland Mining Co.	Md.	54.51	41.09	12.31	73.50	8.18	9.18	8.18	3.09	16.0	9,997
Neff's Cumberland	Md.	53.70	41.71	12.67	74.53	12.40	9.78	12.71	5.43	10.1	11,208
Easby's "coal in store"	Md.	54.29	41.26	14.98	76.26	10.34	9.44	10.96	4.53	6.1	10,604
Atkinson and Templeman's	Md.	53.47	41.90	14.93	76.69	8.08	10.22	8.38	1.33	18.2	10,935
Easby and Smith's	Md.	52.92	42.33	15.53	74.29	7.33	10.70	7.96	3.04	5.1	11,624
Cumberland (Navy Yard)	Md.	51.16	43.78	15.52	70.85	9.30	9.96	9.69	2.12	5.3	11,034
Dauphin and Susquehanna		53.29	42.04	14.87	74.29	14.98		14.53	2.29	13.5	
Blossburgh	Pa	50.54	44.32	13.82	70.85	11.49	9.34	16.36	3.50	23.7	11,171
Lycoming Creek	Pa.	53.05	42.22	14.78	73.11	10.77	9.72	11.20	3.40	13.7	10,956
	Pa.	55.38	40.45	13.84	71.53	13.96	8.91	16.92	3.26	46.2	10,724

TABLE SHOWING THE VALUE AND PROPERTIES OF VARIOUS KINDS OF COAL.—Continued.

DESIGNATION OF COAL		Weight per cubic foot by experiment.	Cubic feet of space required to stow 1 ton.	Volatile combustible matter in 100 parts.	Fixed carbon in 100 parts.	Earthy matter in 100 parts.	Pounds of steam to 1 of coal from 212°.	Total waste (ashes and clinker) from 100 of coal.	Weight of clinker alone from 100 of coal.	Av. weight in pounds of unburnt coke left on grate after such experiment.	Steam from 212° from 1 of combustible matter.
Quin's Run	Pa.	50.34	44.50	17.97	72.79	8.41	10.27	8.94	1.31	14.7	11,275
Karthaus	Pa.	52.54	42.63	19.53	73.77	7.00	9.09	7.89	3.66	12.8	9,887
Cumbria County	Pa.	53.46	41.90	20.52	69.37	9.15	9.24	9.75	3.48	14.8	10,239
Barr's Deep Rum	Va.	55.17	42.13	19.78	67.96	10.49	9.02	11.07	4.78	6.4	10,142
Crouch and Snead's	Va.	53.59	41.80	24.38	59.98	14.28	8.34	14.34	5.37	6.0	9,740
Midlothian (900-foot shaft)	Va.	50.52	44.34	27.38	61.08	10.47	8.58	10.70	6.47	5.9	9,611
Creek Company's Coal	Va.	46.50	48.17	32.47	60.30	8.57	8.42	8.64	4.41	10.5	9,211
Clover Hill	Va.	45.49	49.25	32.21	56.83	10.13	7.07	8.60	3.86	11.5	8,588
Chesterfield Mining Co's	Va.	45.55	49.18	32.63	58.79	8.63	9.00	9.07	4.19	10.5	9,896
Midlothian (average)	Va.	54.04	41.45	29.86	53.01	14.74	8.29	14.83	8.82	6.4	8,741
Tippecanoe	Va.	45.10	49.67	34.54	54.62	9.37	7.75	9.72	4.03	17.1	8,583
Midlothian (new shaft)	Va.	47.90	49.76	35.17	56.40	9.44	8.75	10.26	4.21	14.8	8,751
Midlothian (screened)	Va.	45.72	48.99	34.70	54.06	9.66	8.94	10.27	3.33	43.2	9,970
Midlothian (Navy Yard)	Va.	56.11	41.13	29.12	56.11	14.14			4.42	5.7	
Picton (from New York)	N.S.	53.55	41.83	27.83	56.98	13.39	8.41	13.37	6.13	5.9	8,497
Sidney	N.S.	47.44	47.22	23.81	67.57	5.49	7.99	6.01	2.24	3.7	6,648
Pictou (Cunard's)	N.S.	49.25	45.48	25.97	60.74	12.51	8.48	12.06	6.19	11.1	8,255
Liverpool	Eng.	47.88	46.78	39.96	54.90	4.62	7.48	5.04	1.86	10.7	7,779
Newcastle	Eng.	50.82	44.08	35.83	48.81	9.34	8.66	5.68	5.63	5.7	9,178
Scotch	Scotland	51.09	43.84	39.19	54.93	7.07	6.95	8.25	3.14	9.9	8,942
Pittsburgh	Pa.	46.81	47.85	36.76	58.44	4.97	8.20	5.12	0.94	6.4	7,734
Cannelton	Ind.	47.65	47.01	33.99			7.34	5.12	1.64		7,734
Dry Pinewood		21.01	106.02			0.30	4.69	0.30			4,707

GUIDE POSTS ON THE ENGINEER'S JOURNEY.

TABLE OF RESULTS OF TESTS OF COAL—MADE JUNE TO OCTOBER, 1883.*

NAME OF COAL.	BY WHOM FURNISHED.	Average under evaporation per pound of coal feed water 212° Fahr.	Taking Grape Creek, Illinois, as the basis of comparison at 100, other coal would stand in comparison at
Grape Creek, Illinois...............	Grape Creek Coal Co.........	7.19	100
Wilmington & Springfield, Illinois...	....	7.34	102
Streator, Illinois...................	Arnold & Co.................	7.55	105
Streator, Illinois...................	Chicago, Wilmington, & V. Coal Co...................	7.63	106
Wilmington, Illinois................	....	7.88	109
Indiana Block.....................	E. T. Ellicott...............	7.87	109
Indiana Block.....................	C. B. Niblock...............	8.12	113
Gartsherrie, Indiana Block.........	Watson & Co................	8.52	118
"Kincaid," Hocking Valley........	Baker & Co..................	8.48	117
B. & O. Hocking Valley...........	Moody & Co.................	8.20	114
Briar Ridge, Hocking Valley.......	Shipman.....................	8.60	119
Shawnee, Hocking Valley..........	Dewey & Co.................	8.62	119
Briar Ridge, Hocking Valley (2d test)	Shipman.....................	8.71	121
Laurel Hill, Pittsburgh.............	W. P. Rend & Co............	12.92	179
TESTS OF COAL—MADE APRIL, MAY, AND JUNE, 1885.			
Hocking Valley.....................	W. C. Wyman...............	8.34	116
Jacksonville, Hocking Valley.......	W. P. Rend & Co............	8.78	122
Youghiogheny, Pittsburgh.........	Weaver, Daniels & Co.......	10.40	144
Youghiogheny, Pittsburgh (2d test)..	Weaver, Daniels & Co.......	12.24	170
Osceola, Pittsburgh................	Costello & Co................	12.36	172
Laurel Hill, Pittsburgh.............	W. P. Rend & Co............	13.14	182

* Experiments made by the Chief Engineer and Commissioner of Public Works of the City of Chicago, as published in the *American Engineer*.

In completeness of combustion the semi-bituminous coals are superior to the anthracites.

Fuel is often wasted from the bars not being properly spaced, and the following table may be of service:

AIR SPACES BETWEEN GRATE BARS.

Lehigh anthracite, pea coal	$\frac{1}{4}$ of an inch.
Schuylkill anthracite, pea coal...	$\frac{3}{8}$ "
Lehigh anthracite, chestnut coal...	$\frac{3}{8}$ "
Lehigh anthracite, stove coal...	$\frac{1}{2}$ "
Lehigh anthracite, broken coal	$\frac{5}{8}$ "
Cumberland bituminous coal	$\frac{3}{4}$ "
Wood	$\frac{3}{4}$ to 1 "
Sawdust	$\frac{1}{16}$ to $\frac{1}{4}$ "

NOTE.—In estimating for a consumption of 14 lbs. of coal per square foot of grate per hour, about 8 lbs. of water may be taken as the rate of evaporation per 1 lb. of coal, which can be done with a good natural draught. With a forced draught and 28 lbs. of coal to the square foot of grate, the rate of evaporation is only about 6 lbs. of water to 1 lb. of coal.

The semi-bituminous coals occupy rather the smallest space per ton weight (42.0372 cubic feet), the anthracite ranking next (42.13 cubic feet), the bituminous coals of Pennsylvania ranking third (42.671 cubic feet), and next the coking coals of Virginia, being the only free burning coals which are decidedly lighter (45.8804 cubic feet), indicating that anthracite is the heaviest class of coal.

In buying anthracite coal, that quality should be selected which has a conchoidal fracture and a bright appearance. If it is of a dull appearance and shows seams and cracks, it will fly into fragments in the furnace, and will not prove economical.

With bituminous coal, if the fractures present a whitish film or rusty stains, they are indications of sulphur and pyrites, and such coal should be rejected for furnace use.

Cumberland coal appears to give the best satisfaction generally, as it is easily handled, there is less waste, and it kindles quickly, while, in its evaporative power, it is superior to anthracite.

The usual waste from anthracite coal may be taken as $16\frac{1}{2}$ per centum, while that from Cumberland coal is only about half so much.

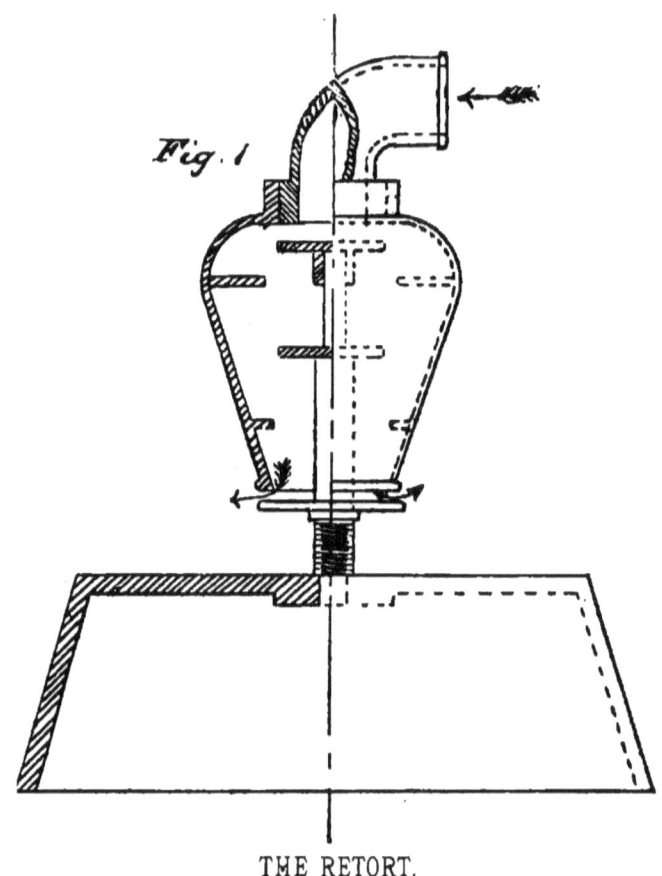

THE RETORT.

The following table presents a comparison of coals and lignites from different sections of this country:

COAL.		PERCENTUM OF ASHES.	THEORETICAL VALUE.	
Kind.	Locality.		Heat (units) per pound.	Lbs. of water evaporated from and at 212° F.
Anthracite	Pennsylvania	3.49	14,199	14.70
"	"	6.13	13,534	14.01
"	"	2.90	14,221	14.72
Cannel	"	15.02	13,143	13.60
Connellsville	"	6.50	13,368	13.84
Semi-bituminous	"	10.77	13,155	13.62
Stone's Gas	"	5.00	14,021	14.51
Youghiogheny	"	5.60	14,265	14.76
Brown	"	9.50	12,324	12.75
Caking	Kentucky	2.75	14,391	14.89
Cannel	"	2.00	15,198	16.76
"	"	14.80	13,360	13.84
Lignite	"	7.00	9,326	9.65
Bureau Co	Illinois	5.20	13,025	13.48
Mercer Co	"	5.60	13,123	13.58
Montauk	"	5.50	12,659	13.10
Block	Indiana	2.50	13,588	14.38
Caking	"	5.66	14,146	14.64
Cannel	"	6.00	13,097	13.56
Cumberland	Maryland	13.98	12,226	12.65
Lignite	Arkansas	5.00	9,215	9.54
"	Colorado	9.25	13,562	14.04
"	"	4.50	13,866	14.35
"	Texas	4.50	12,962	13.41
"	Washington Territory	3.40	11,551	11.96
Petroleum	Pennsylvania	....	20,746	21.47

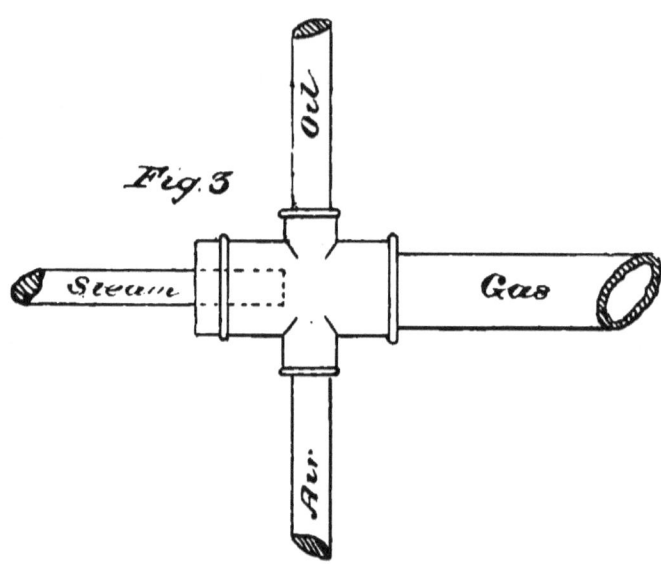

THE ATOMIZER.

GUIDE POSTS ON THE ENGINEER'S JOURNEY.

Attempts have been made, at various times during the past twenty-five years, to make use of petroleum or rock oil as a substitute for coal under steam boilers, and with varying success. Its evaporative power from and at 212° Fahr. is 20.23, while the best practical results obtained from coal do not exceed 11.6. Now, as an actual fact, an evaporation in regular working of 18 has been obtained from petroleum, and the usual results are a little over 16—while the results generally obtained from coal are below 7—so that it seems, at first sight, as if coal were about to be superseded, especially as by its use fewer men are required; and the trouble of sweeping flues and removing soot and ashes is avoided. But here the question of cost of the oil comes in to determine the practicability of its use, and that has hitherto prevented its employment to any great extent, except in certain sections, notwithstanding the fact that a quantity of oil, equal in practical evaporative effect to one ton of coal, can be stowed in one-third of the space.

In Russia, where coal is high and petroleum not only cheap but plentiful, it has been used very successfully, and with a great reduction of expense, upon the locomotives of various railways, and to some extent upon steam vessels. In this country, in Cleveland and other places west of the Alleghanies, it is also used with a great saving of expense; and latterly in California it has been used on a few steamers.

There are two methods of burning petroleum—one is the retort, and the other the atomizing process. There are many patentees of apparatus for burning oil, and their devices vary considerably in detail, but they all can be classed under the head of one or the other of the two processes mentioned.

In the accompanying figures is shown an apparatus of the "Retort Burner" class, which is meeting with much favor in the neighborhood of Toledo, Ohio. The following is its mode of operation:

The tank containing the oil is placed at some distance from the boiler, and at a height of two or three feet above the retort, Fig. 1. The part of the apparatus situated outside the furnace is shown in Fig. 3. The oil is fed from the tank into one arm of the cross by gravity; is there met by a jet of steam and driven through the pipe, marked gas, into the top of the retort in the furnace, passing out at the bottom, where it is deflected by an adjustable plate. This retort rests on a hollow base. To place the retort, two or three grate bars are removed from the furnace; the base is then set over the opening, and then the grate is solidly bricked over close up to the base on all sides, so that no air can get into the furnace except through the opening in the base. At a distance from the base of about eight or ten inches, an open loose wall of brick, without mortar, is laid up concentric with it, no two bricks being allowed to approach each other at the ends within two or three inches. The fire is started by letting a little oil drop through a separate pipe upon a plate of asbestos underneath the base, the cocks on the oil and air pipes, Fig. 3, being slightly opened so that, as the retort gets heated, a little gas is made. As soon as steam is raised on the boiler the oil is shut off from the starting pipe; the steam is turned on in the jet; the air-cock is shut, and the oil regulated to suit the amount of fire required. An evaporation of from 14 to 16 lbs. of water by 1 lb. of oil is claimed for this apparatus.

There is no danger of burning or cracking the plates of the boiler with this apparatus, as there is no direct impact of flame.

There is, however, a strong probability of cracking the retort, and also of carbon caking in the retort, thus clogging it and reducing its efficiency.

In the Acid Works at Cleveland, Ohio, the residuum from the oil refineries is treated to recover the acid, and the refuse that remains is used to burn under the stills and to generate steam.

The burner used under the boilers is of the atomizing type, and was invented by the assistant superintendent of the works, and is economical, as one and one-half to two barrels of the refuse only will evaporate as much water as one ton of coal.

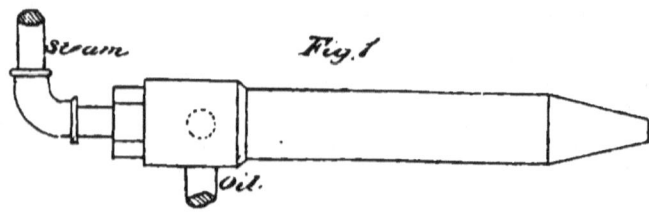

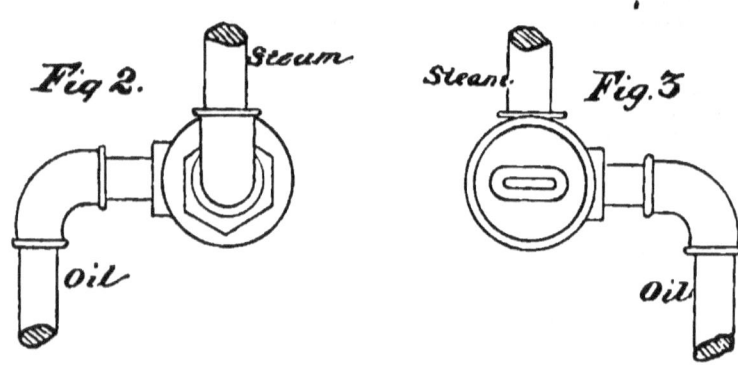

In the plates, page 72, Fig. 1 shows a side elevation of the apparatus; Figs. 2 and 3 show back and front end views of the same; Fig. 4 is a vertical longitudinal section of the outer shell; and Fig. 5 is a similar view of the inner pipe. This apparatus is made wholly o cast brass, and is operated as follows:

The pipe shown in Fig. 5 is screwed into the shell shown in Fig. 4, leaving a concentric chamber between them; a suitable pipe for the conveyance of steam is connected to the large end of this inner pipe; and a pipe for the oil conduit is connected to the hole shown in the side of the large end of the shell. The furnace is prepared by taking out the grates and bricking up the furnace door, 'eaving a hole in the brick-work at the bottom to allow the nozzle of the apparatus to project a few inches into the furnace. A quantity of old bricks, also, are thrown into the furnace against the bridge wall, roughly sloping toward the ashpit door.

The oil is supplied to the apparatus, by gravity, from a small tank situated at some distance from the boilers, and at a height of from three to five feet above the apparatus.

The steam is supplied from the boiler direct, as it is essential to have the pressure constant, and if taken from the main steam pipe it would be subject to pulsations from the engine motion.

The action is as follows:

On opening the oil cock, the oil drops on the inner pipe, and is heated by opening the steam valve and permitting the steam to rush through; it is also partially vaporized, and is drawn rapidly through the annular chamber by the vacuum formed by the steam escaping at the flattened mouth of the apparatus, becoming warmer as it progresses, until it strikes and mingles with the steam, where it is almost wholly atomized and converted into vapor, or gas, and thrown forcibly into the furnace, where it is kindled by a few burning shavings.

It has been found that if the oil tank is supplied with a steam coil, and the oil warmed to a temperature of from 130° to 180° Fahr., much better results are obtained than when the oil enters the apparatus cool.

The steam pipe is generally one-quarter inch in diameter, and the oil pipe one-half or three-quarters of an inch.

At Messrs. Scofield, Sherman & Teagle's oil refinery in Cleveland, O., an oil from Findlay, O., is used as fuel under the boilers and stills. They make use of the Edwards burner, also of the atomizing type, which is shown in Figs. 1 and 2, page 74, being respectively an elevation and a longitudinal vertical section.

In this apparatus the oil is supplied by gravity to an internal tube, the mouth of which can be partially or entirely closed by a conical enlargement on the end of a hollow rod extending through the cylindrical oil chamber and tube, and which is supplied with a hand-wheel and screw-thread to give it end motion, and thus regulate the supply to the furnace.

This oil chamber and tube is surrounded by a casing, into which steam is admitted, and it is contracted at its end so as to allow of a small opening concentric with the oil pipe; thus the steam warms and vaporizes the oil and forces it into the furnace. An outer concentric shell surrounds part of the steam casing, which supplies air around the jet of oil gas, while the hollow rod, regulating the oil supply, furnishes air to the interior of the gas jet.

This apparatus is very economical, as it is found that one and a half barrels of the Findlay oil, when burned in it, will produce the same amount of evaporation as one ton of coal.

It should be remarked, however, that it requires a pressure of about 20 lbs. of steam in order to operate successfully either this apparatus or the one preceding.

The Lima oil, as well as that from several other districts, can not be refined, and therefore can only be utilized as a fuel. Its price at the well is only 20 cents per barrel, and therefore it is more economical than coal within a large radius, the distance depending on the freight charges mainly.

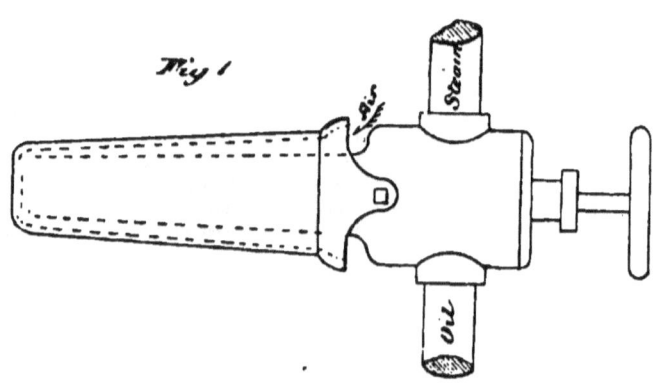

Fig 1

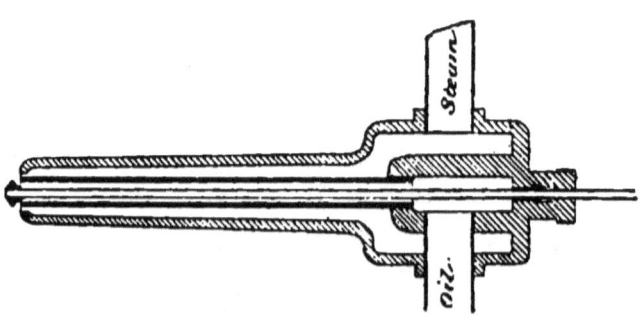

The Cleveland Refining Co., of Cleveland, O., make use of a simpler apparatus, also of the atomizing variety.

Fig. 1, shown below, is a front elevation of the apparatus as applied to the boiler; *a* is the steam pipe, and *b* is the oil pipe. The furnace doors are removed and the opening bricked up, with the exception that one brick is removable for a sight-hole. The ashpit openings are bricked up, as shown, to the springing of the arched tops, and the apparatus is placed in the opening thus left. The arrangement of the brick bed in the furnace is shown in Fig. 2, page 76. The vertical elevation of the apparatus is shown in Fig. 3, while a plan of it is given in Fig. 4, the dotted lines show-

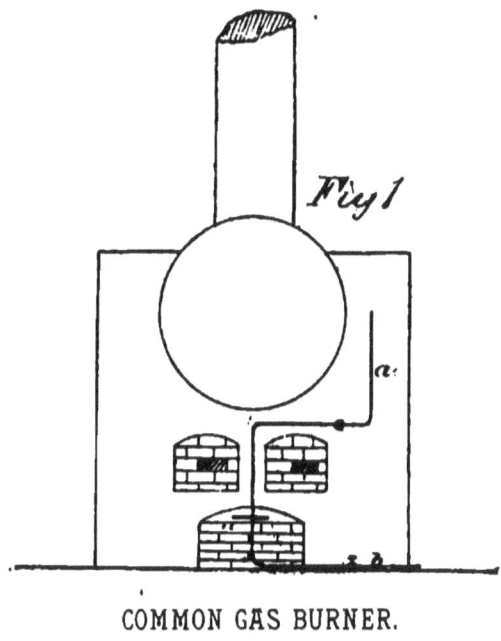

COMMON GAS BURNER.

ing the arrangement of the internal pipes. The oil pipe enters at the external end, being supplied by gravity. The steam enters the apparatus and surrounds the oil pipe, warming the oil, and terminates at the internal air openings, drawing warm air from the front of the furnace chamber, and driving it through the jet pipe into the furnace, together with the partially atomized and vaporized oil, where the combination makes heat with much flame.

With this apparatus, thus applied, it is claimed that two barrels of oil are equivalent in evaporative power to one ton of coal. It is easily managed, no smoke is made, and the flame does not strike the sheets of the boiler so as to produce dangerous effects.

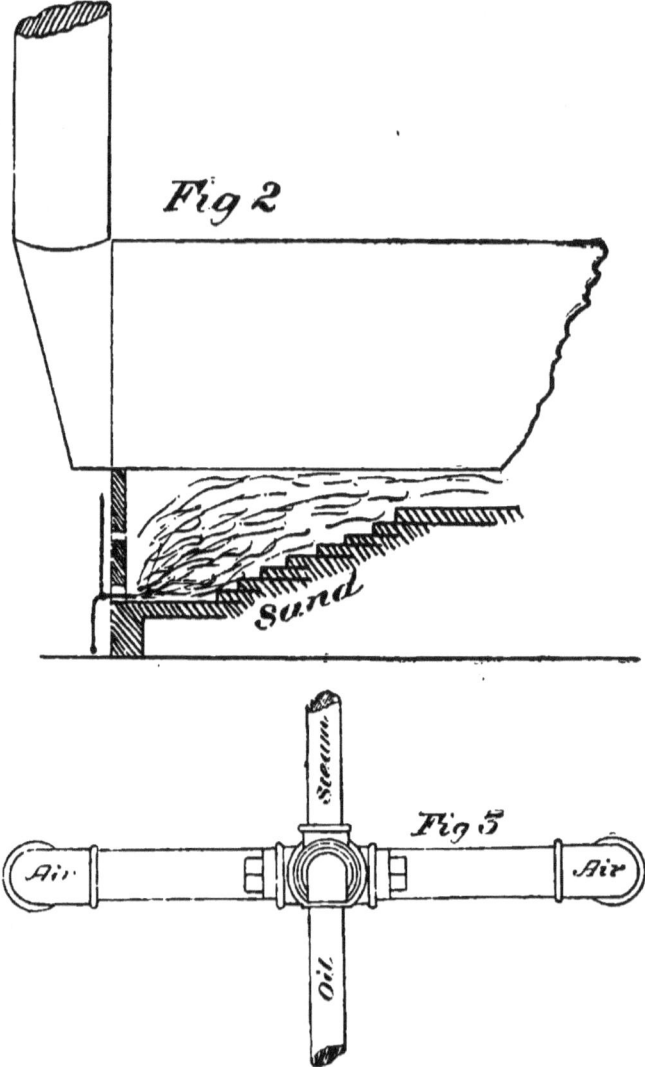

THE ATOMIZER.

At the Britton Iron and Steel Co's Works, Cleveland, O., another method of burning oil is used, also of he atomizing variety.

Fig. 1, page 78, is an end view of two boilers, with one front removed in order to show the internal arrangement of the furnace, which was formerly used to burn coal.. The oil pipe extends across the boiler fronts above the furnace doors, with drop branches to the apparatus. The oil is

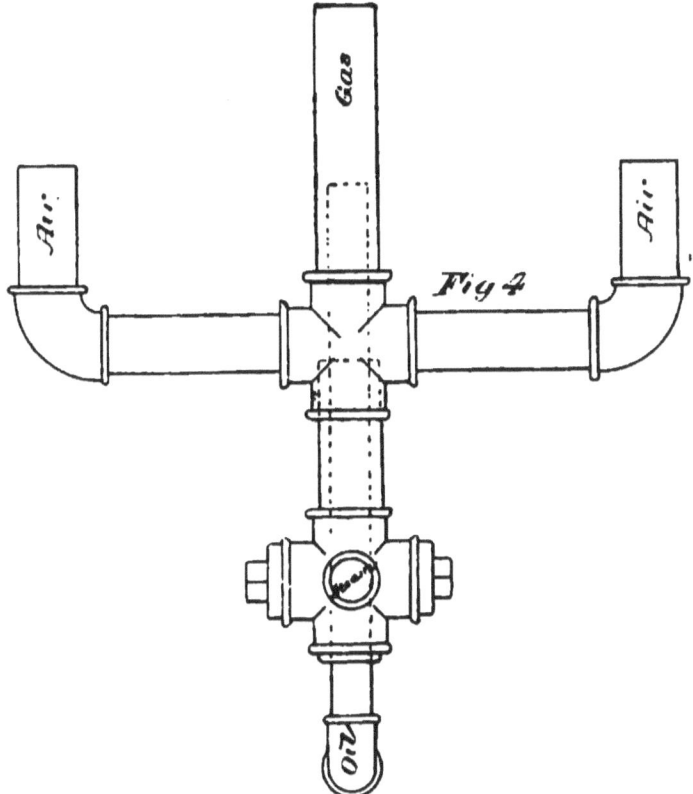

ATOMIZER FOR USING HEATED AIR.

supplied from a tank outside the building, and flows to the apparatus by gravity. The steam pipe enters the furnace for a distance of about three feet, and returns outside again before being connected to the apparatus, forming a close U; this is for the purpose of drying and superheating the steam, which enters the apparatus at two points. Where it enters at the rear it simply induces a current of air; at the side entrance it atomizes the oil.

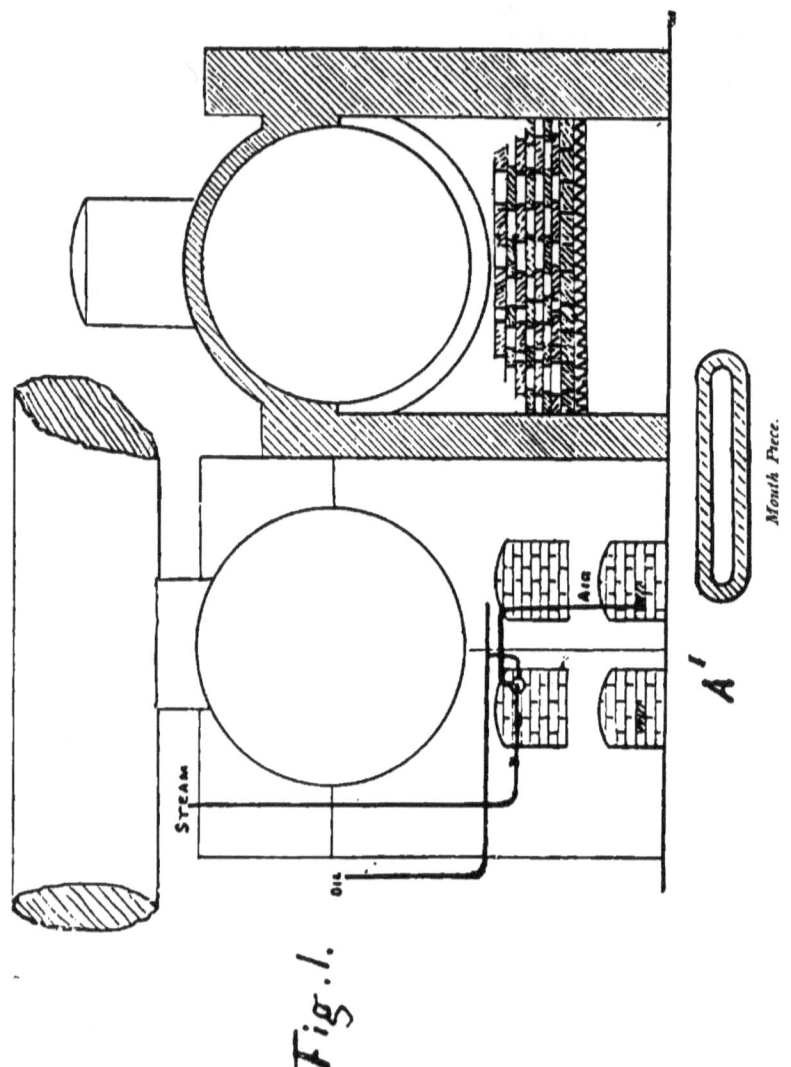

Fig. 1.

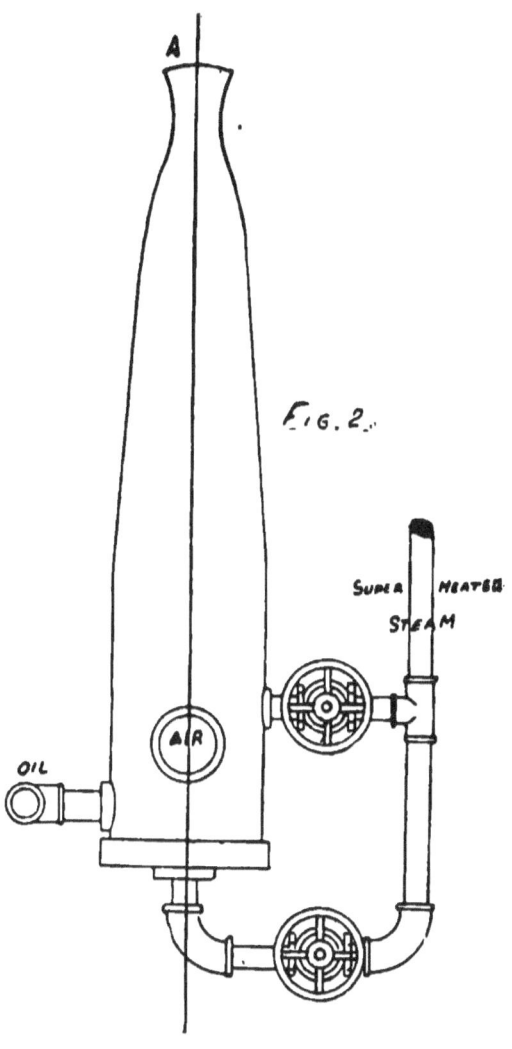

Fig. 2.

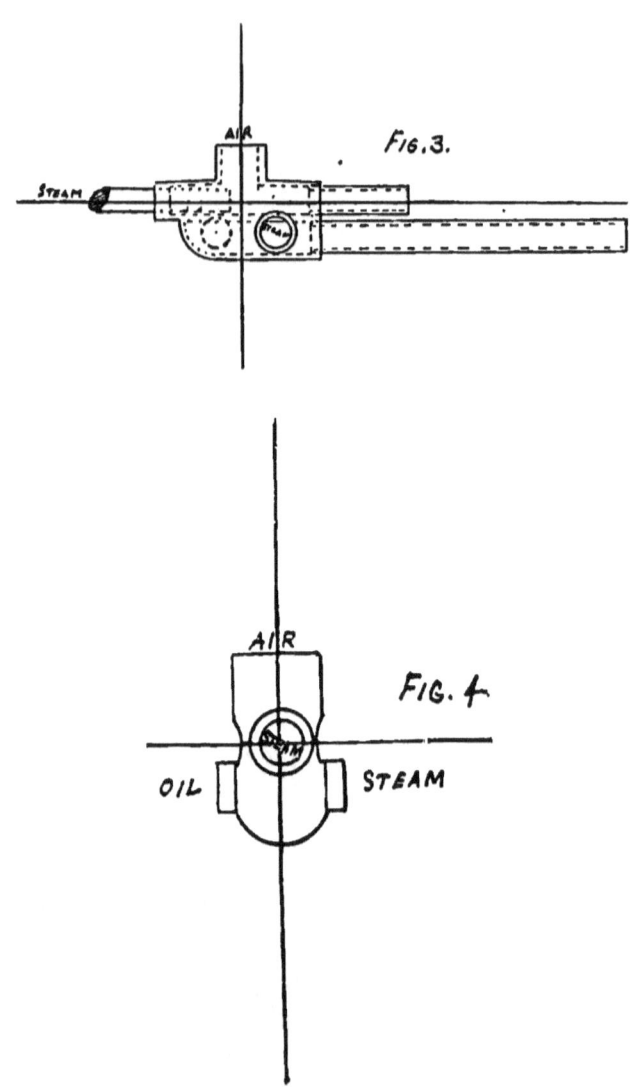

The air pipe is led from the ashpit for the purpose of securing a supply of warm air, the ashpit being arranged act as a hot-air chamber. This arrangement is, however, worse than useless, as has been demonstrated in many instances, a greater economy having been always attained where the air has been supplied at a normal temperature.

Fig. 2, page 79, is a view of the burner and its connections, looking down on it when in position.

In Fig. 3 is shown a side elevation of the apparatus with the external shell removed.

Fig. 4 shows an end elevation of the internal apparatus.

It will be seen, in Fig. 1, that the grate bars have been bricked over, forming an air-tight bottom to the furnace, and also that an open brick wall extends across the furnace. This wall is n front of the bridge, and at a distance of about three feet from the furnace front; its use is two-

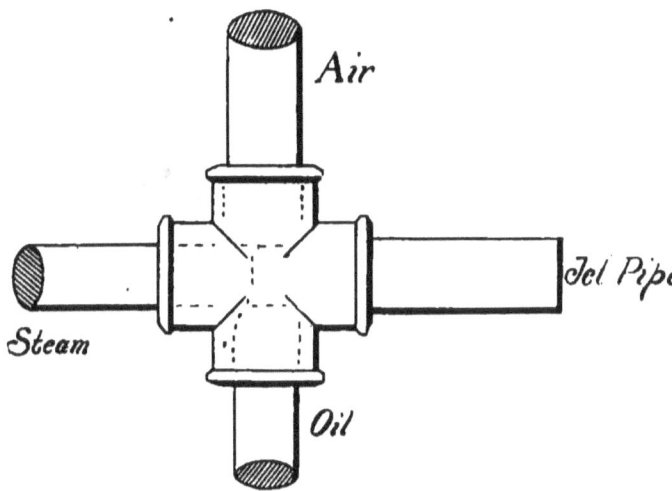

old—first, or the purpose ot spreading and dividing the flame; second, to retain more of the heat at the front end of the boiler.

The mouth piece of the burner, as shown at A, in Fig. 2, page 79, is flattened, and the orifice is shaped as shown in A^1, page 78, in order to spread the flame; but this variation is of doubtful utility, as it is very contracted; some claim that it hinders vaporization.

The most simple of the different varieties of the atomizing style is given in the above figure. It consists of a simple fitting, a cross; it is placed in an upright position, with the pipes connected to it, as shown in the figure. The oil is supplied to it by gravity from a tank, the steam entering at the rear atomizes the oil, induces a current of air, and acts as a mixer, while warming both air and oil, and throws them into the furnace with considerable force. This apparatus will work well with only five pounds steam pressure, though it is said to be not quite so economical as the others that have been mentioned. But it possesses the advantages of being cheaply and easily made,

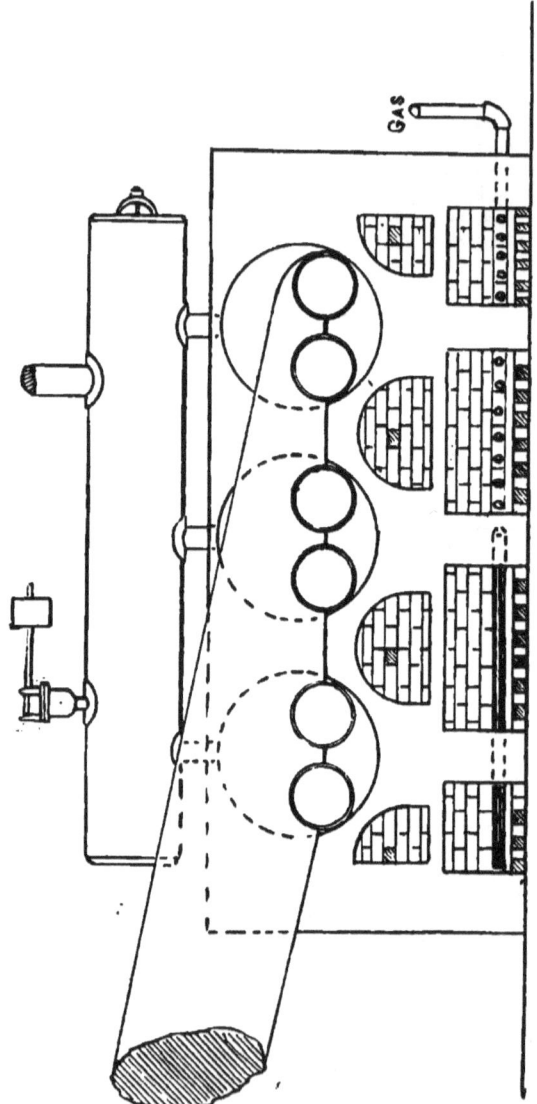

Fig. 1.

ARRANGEMENT OF PIPES FOR ORDINARY BURNER.

and with no liability to get out of order. The steam pipe is generally one-half inch in diameter, the oil pipe three-quarters of an inch, and the air pipe an inch. The jet is generally inch pipe, and not much above six inches in length.

Natural gas within a few years has attained a prominence as fuel, especially at and around Pittsburg and other Western cities. It burns without odor or smoke, the amount of heat can be regulated at will, and it is cheap—all valuable qualities.

Figs. 1 and 2 show a battery of three double-flue boilers. The doors of the furnaces are bricked up, allowing one brick to be loose for a sight-hole, the ashpit is floored tightly with tiles, supported by bricks laid on edge, to within about fifteen inches of the bridge wall, and from the bridge wall towards the furnace front, over the other tiling, and at about four inches from it, is

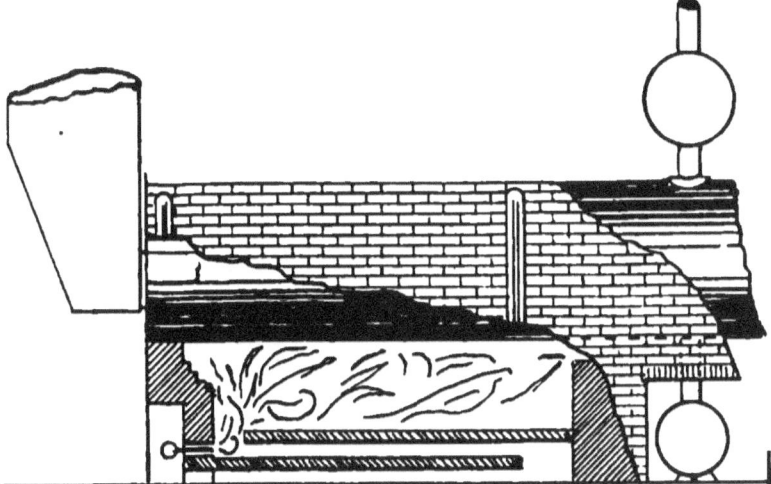

FIG. 2.

laid another floor of tiling. As the ashpit doors are bricked up, the only entrance for air into the furnace is to flow in between the rows of supporting brick towards the bridge wall, and from thence, between the two floors of tiling, to the front of the furnace, where it abuts against the streams of entering gas, thoroughly mixing. This method is one of the oldest methods known for burning gas, but it is far from being the best, though it is commonly used.

Fig. 2, page 84. Here the furnace bars are dropped, and they are bricked over to within eighteen inches of the bridge, and then a tile flooring laid about four inches above the brick, from the bridge to within a foot of the front of the furnace. The gas pipe enters the furnace and is furnished with a short cross pipe in which are a number of short vertical jet pipes. The air meets the entering gas as before, and the mixture burns freely without smoke or odor. This is about as economical a method as the other; there is little to choose between them.

Fig. 3, page 85, shows a method of applying gas burning to a pair of boilers, so that a coal fire can be made use of without much trouble, in case of failure in the gas supply.

Fig. 2.

ANOTHER FORM OF ORDINARY BURNER.

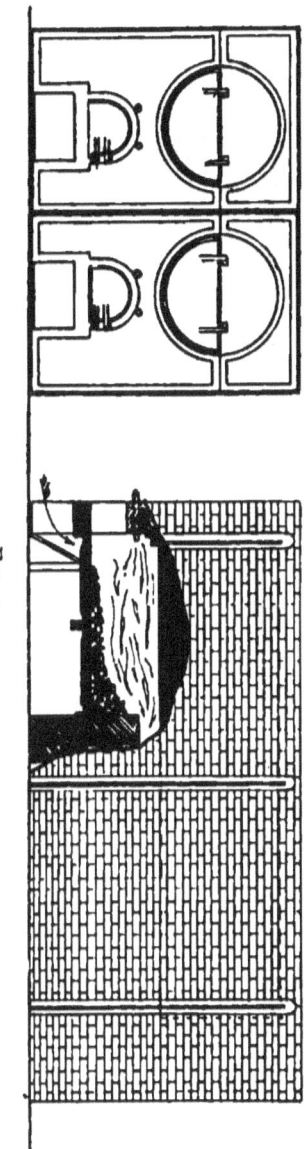

Fig. 3.

A LOW PRESSURE BURNER.

86 GUIDE POSTS ON THE ENGINEER'S JOURNEY.

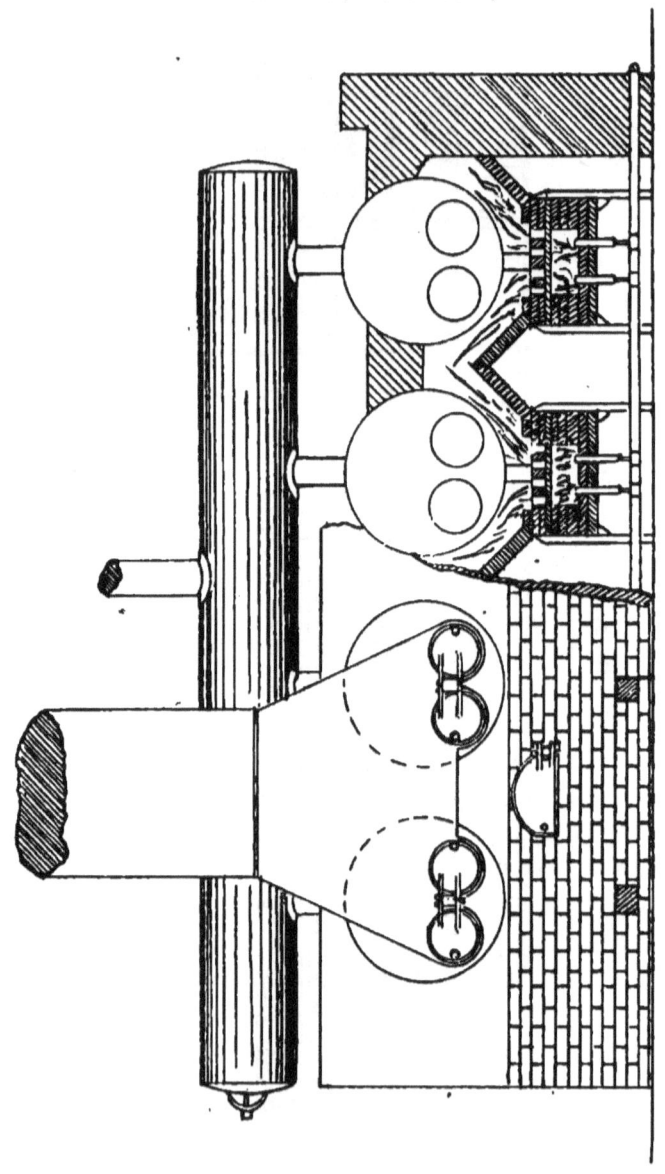

BEST FORM OF BURNER.

The gas is introduced in two separate jets through pipes over each furnace door; the grates are covered with a thick layer of cinders and ashes, to within a foot or so of the furnace doors; the ashpit doors are removed entirely, and boards are placed slanting, from the ashpit doors up to the grate, so as to direct a current of air through the bars at the front end. The pressure of the gas is low where this apparatus is used, and occasionally fails entirely, as the mains from the well are small. The size of the jet pipes is about 1¼ inches.

In Fig. 4, on this page, is shown another method in use where the pressure of the gas is low, and liable to fail. This apparatus consists of a pipe entering the furnace at the level of the grates, lying upon them, and running about two-thirds of the length of the furnace. The back ends of the grates only, in this case, are covered with ashes and cinders. The ashpit doors are removed, but no other change is made. This gas pipe has a diameter of 1½ inches, and is perforated with three rows of holes on its upper side.

In the figures on pages 86, 88, and 90 is given the method of burning gas adopted by Singer, Nimick & Co., Pittsburg, Pa., and it is by far the most economical. The drawing on page 86

FIG. 4.

LOW PRESSURE BURNER.

shows a battery of four ordinary flue-boilers, the front being broken away from two of them to show the furnace arrangements. There are three furnaces, or collections of gas jets, under each boiler, to equalize the temperature. Each furnace, practically, consists of a box made of fire-brick, supported on an iron frame. The top of the box consists of tiling and ordinary fire-brick, arranged to leave plenty of openings, and tiles are so arranged between the furnaces as to retain the heat and reflect it upon the boiler. The burners of each furnace are really Bunsen burners, and they are arranged longitudinally of the boilers in two rows, each athwartship row of furnaces being supplied by a separate main. These burners just extend through the brick bottom of the furnace, and no more. At night, when but little is demanded from the boilers, it is the usual custom to extinguish two of the athwartship rows of furnaces, alternating every night, to save undue strains from unequal expansion and contraction. This method is the invention of the superintending engineer of the works. It is very efficient, and it seems to be very durable, requiring little in the way of renewal or repairs.

THE BURNER.

There is much said lately of a falling off in the yield of natural gas wells, and, though new ones continue to be discovered, there appear to be grounds for fear that they may fail entirely. Under these circumstances the probability that petroleum, or petroleum gas, will be the fuel of the future becomes stronger. Therefore it is of no common interest to both the owners and the engineers of steam plants to be acquainted with the various devices for utilizing that material as fuel. It is useless to attempt, in a work of this extent, to enumerate all of the many and various devices that have been invented for the purpose. Some few of the principal ones have been shown and described, but their number is continually increasing, and since this article has been written, one of the very best burners for the use of natural oil, or refuse, called the Eclipse, has been placed upon the market, and has superseded other burners of the highest grade hitherto known. This new burner is neither a retort burner nor an atomizer simply—but a combination of the best features of both. It is the invention of an engineer connected with the Standard Oil Company, at Cleveland, O., and is the result of long study and careful experiment on a practical commercial scale. It has already gained for itself a reputation second to none, and possesses many advantages over all others, chief among which is the facility with which it can be removed from a furnace, in case of failure from any cause, and the short time required for preparing the furnace for a coal fire—there is required for the whole operation of pulling out the burner and starting a coal fire not exceeding ten minutes. The apparatus can be easily operated, and the usual evaporation has been demonstrated to be 18 pounds of water per 1 pound of coal under ordinary circumstances, collating the results of several weeks' actual use, and higher results have been obtained on short experimental runs.

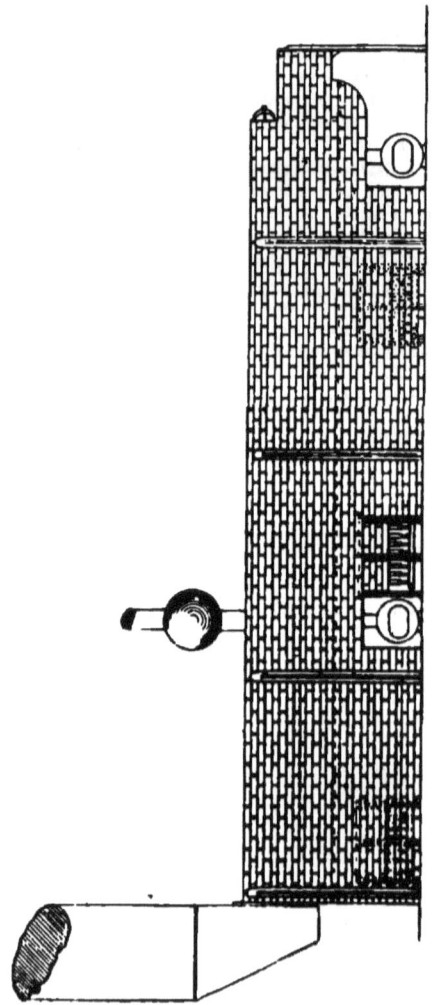

ANOTHER VIEW OF THE APPARATUS.

COMBUSTION.

Without perfect combustion a great portion of the heat which the uel is capable of yielding is not developed, and hence loss results. If the surfaces are not properly adapted for the reception of the heated given out by the fuel, the gases pass into the chimney at too high a temperature, with loss again. Heating the feed water by waste steam, or waste gases, leaves less to be done by the fuel in the furnace, and by introducing the feed near the water-level less contraction and expansion of the plates occur. And if a boiler is so constructed that the steam and water currents in ascending and descending interfere with each other, or if the configuration of the parts obstructs their proper motion, foaming, overheating, burning of the plates, etc., result, causing great expense for repairs.

Combustion is an energetic chemical combination of oxygen with some substance, accompanied by light and heat. The substance with which it combines is called the combustible, or, when the combustion takes place in a furnace, stove, or grate in the ordinary use of every-day life, it is called fuel, as for instance, wood, charcoal, coke, coal, oil, etc.

The oxygen is supplied from the atmosphere by which we are surrounded. Air consists of two gases, nitrogen and oxygen, of which the oxygen forms only one-fifth part in bulk, the gases not being in a state of chemical combination, but forming a mixture only.

In the flame of a common candle, combustion takes place in the following manner:

The wick is filled with wax, tallow, or other oleaginous material, and on a match being applied this material is converted into gases, which ignite, forming flame. This flame is divided into four distinct portions, or coatings, differing both in their aspect and the nature of the processes producing them. Immediately surrounding the wick is a dark space (2) of a conical shape, filled with the combustible gas unconsumed, and continually generating by the first action of the heat upon the fuel (wax, tallow, etc.).

Surrounding the base of the dark cone and the lower portion of the luminous part is a cup-shaped cone (1) of a blue tint, faintly luminous, but sharply defined. It results from the sudden and complete combustion of the gases of the dark cone (carbureted hydrogen principally) with a full supply of air (oxygen) striking them from without.

Above the dark cone lies the luminous portion (3) of the flame; the oxygen of the air uniting with the hydrogen, and raising the now separated carbon to the high temperature of incandescence, which gives the luminosity to this portion of the flame.

Exterior to all this is the cone of final and complete combustion (4), in which the highly heated carbon atoms unite with oxygen in a combustion of their own, being converted into carbonic acid. This cone (4) is only faintly luminous, it surrounds the flame on all sides, and is its hottest portion. The maximum temperature is a little *above* the point of the luminous cone, where we also find the highest *oxidizing* power, while just *within* the luminous point the high temperature, and the presence of free carbon co-operate to produce the most energetic *reducing* action.

The products of perfect combustion are thus shown to be water (steam) and carbonic acid, and to insure it, a sufficiently high temperature and a sufficient supply of oxygen are necessary.

The first step towards effecting the combustion of any gas is to ascertain the quality of oxygen with which it will chemically combine, and the quantity af air required to supply that amount of oxygen. Much of the apparent complexity which exists on this head arises from the disproportion between the relative *volumes*, or *bulk*, of the constituent atoms of the several gases, as compared with their respective *weights*. For instance, an atom of *hydrogen* is *double* the bulk of an atom of *carbon vapor;* yet the latter is *six times the weight* of the former.

Again, an atom of hydrogen is *double* the bulk of an atom of oxygen; yet the latter is *eight times the weight* of the former.

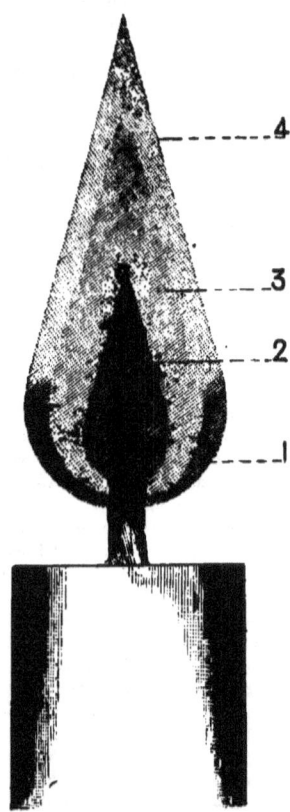

So, of the constitituents of atmospheric air, nitrogen and oxygen, an atom of the former is *double* the bulk of an atom of the latter; yet in weight it is as *fourteen to eight*.

Hydrogen, separating from its combination with carbon carbureted hydrogen) on the application of heat, and uniting with oxygen, produces water in the shape of steam, and this atom of steam occupies a bulk only two-thirds of that of both—as below:

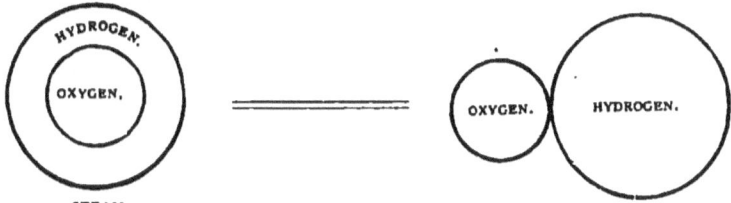

STEAM.

Again, the carbon meeting with its equivalent of oxygen, unites with it, forming carbonic acid gas, composed of *one atom* of carbon (by weight 6) and *two atoms* of oxygen (by weight 16), the latter in volume being double that of the former, as in the annexed figure:

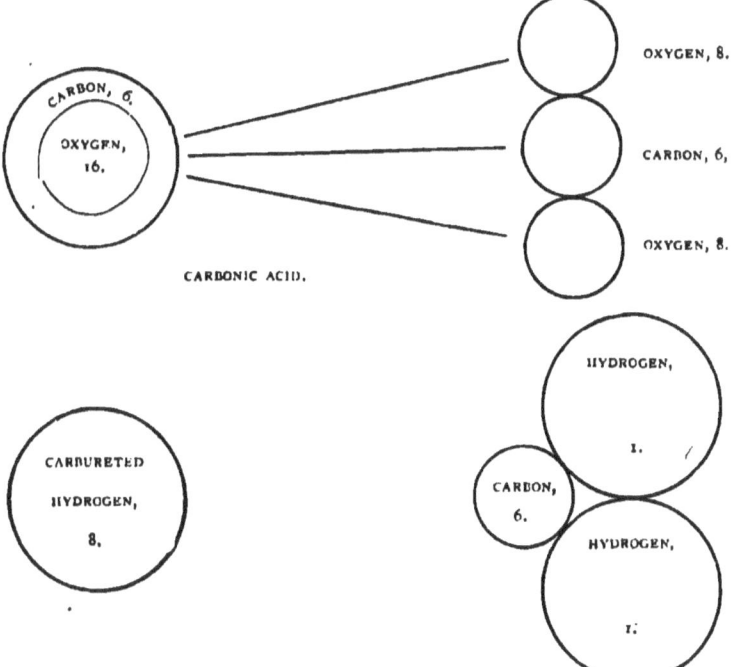

Carbureted hydrogen is composed of *two atoms* of hydrogen (by weight 2) and one atom of carbon vapor (by weight 6), and their resultant bulk is that of one atom of hydrogen only, as in the annexed figure:

94   GUIDE POSTS ON THE ENGINEER'S JOURNEY.

Atmospheric air is composed of two atoms of nitrogen and one atom of oxygen, each of the former being double the volume of an atom of the latter, while their relative weights are as fourteen to eight. But this is only a mechanical mixture, the nitrogen not being in chemical combustion with the oxygen. The gross weight of the nitrogen is to that of the oxygen as twenty-eight is to eight, and the gross volume as four to one, as in the annexed figure:

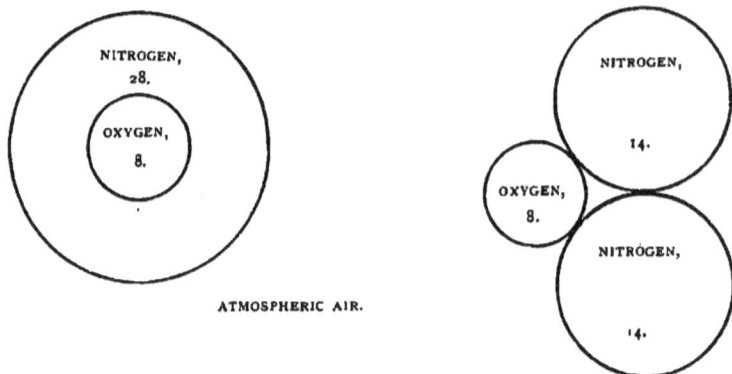

ATMOSPHERIC AIR.

Now, having ascertained the quantity of *oxygen* required for the saturation and combustion of the two constituents of carbureted hydrogen, the remaining point to be decided is the *quantity of air that will be required to supply this quantity of oxygen*.

This is easily determined, as we know precisely the proportion which oxygen bears, in volume, to that of the air. For, as oxygen is but *one-fifth* of the bulk of the air, five volumes of the latter will be necessarily required to produce *one* of the former; and as we want *two* volumes of oxygen for each volume of the gas, it follows that, *to obtain these two volumes, we must provide ten volumes of air*.

The annexed diagram shows the volume of air required for the combustion of the gas.

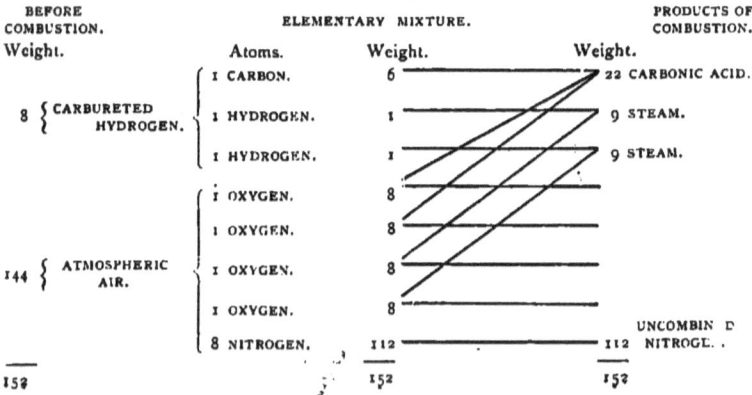

Now, when burning bituminous coal, after the gas has been consumed, there still remains the carbonaceous part, coke, resting upon the bars, to be disposed of.

Carbon is capable of being united with oxygen in two proportions, by which two distinct bodies are formed, possessing distinct chemical qualities.

The proportions in which carbon unites with oxygen form, first, *carbonic acid;* second, *carbonic oxide.* Carbonic *acid*, we have seen, is a compound of one atom of carbon with two atoms of oxygen; while carbonic *oxide* is composed of one atom of carbon and only *half* the quantity of oxygen, as in the annexed figures:

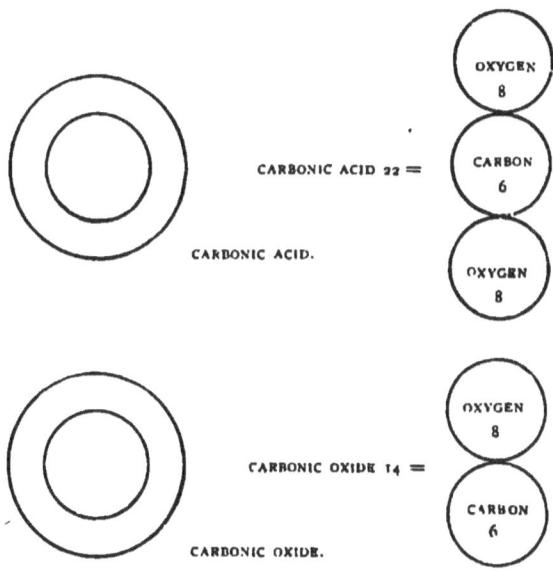

Here we see that carbonic *oxide*, though containing *but half* the quantity of oxygen, is yet of the same volume as carbonic acid, a circumstance of considerable importance on the mere question of *draught* and supply of air, as will be shown hereafter.

Now, the combustion of this *oxide, by its conversion into the acid*, is as distinct an operation as the combustion of the carbureted hydrogen, or any other combustible.

But the most important view of the question is as regards *the formation* of this *oxide;* and this is the part of the inquiry which most requires our attention.

The direct effect of the union of carbon and oxygen is the formation of carbonic *acid*. If, however, we *abstract* one of its portions of oxygen, the remaining proportions would be those of carbonic *oxide*. It is equally clear, however, that if we *add* a second portion of *carbon* to carbonic *acid*, we shall arrive at the same result, namely, the having carbon and oxygen combined in equal proportions, as we see in carbonic *oxide*.

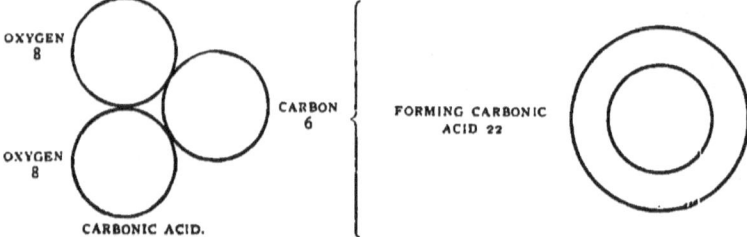

CARBONIC ACID.

By the addition, then, of a second portion of carbon to the above, two volumes of carbonic oxide will be formed, thus:

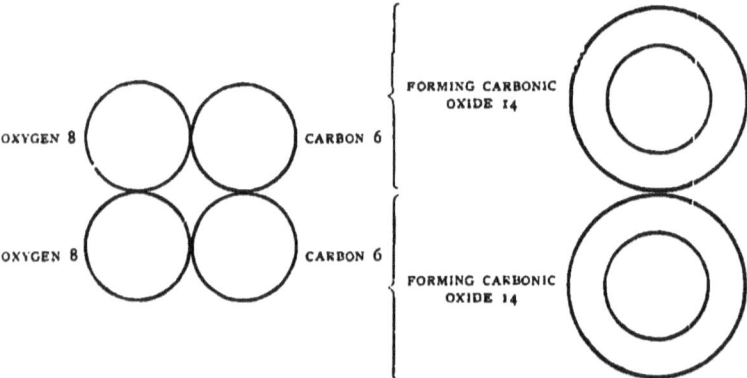

Now, if these two volumes of carbonic oxide can not find the oxygen required to complete their saturating equivalent, they pass away necessarily but *half consumed*, a circumstance which is constantly taking place in all furnaces where the air has to pass through a body of incandescent carbonaceous matter.

The most prevalent operation of the furnace, however, and by which the largest quantity is lost in the shape of carbonic acid, is thus: the air, on entering from the ashpit, gives out its oxygen to the glowing carbon on the bars and generates much heat in the formation of carbonic acid. This acid, necessarily at a very high temperature, passing upwards through the body of incandescent solid matter, takes up an additional portion of the carbon and becomes carbonic *oxide*.

Thus, by the conversion of one volume of acid into two volumes of *oxide*, heat is actually absorbed, while we also lose the portion of carbon taken up during such conversion, and are deceived by imagining that we have "burned the smoke."

Another important peculiarity of this gas (carbonic oxide) is that, by reason of its already possessing *one-half* its equivalent of oxygen, it inflames at a lower temperature (about 600°

GUIDE POSTS ON THE ENGINEER'S JOURNEY. 97

Fahr.) than the ordinary *coal* gas; the consequence of which is that the *latter*, on passing into the flues, is often cooled down below the temperature of ignition; while the *former* is sufficiently heated, even after reaching the top of the chimney, and is there ignited on meeting the air.

We may thus set it down as a certainty that, if the carbon, either of the gas or of the solid mass on the bars, passes away in union with oxygen in any other form or proportion than that of *carbonic acid*, a commensurate loss of heating effect is the result.

Now, in order to kindle a substance and keep it burning, it must be heated to a certain degree, and kept up to that temperature.

The phosphorus of a match inflames so readily that mere friction ignites it, and, in burning, it gives out heat enough to ignite the sulphur of the match, which in turn ignites the wood of the match, and by means of this last flame we ignite the kindlings, and they in turn ignite the larger pieces of wood, and the heat given out raises the temperature of the coal sufficiently for it to ignite; and thus we see that ignition of the coal is the last of a series of a progressive steps—each increasing in temperature.

Smoke is a sure evidence of imperfect combustion; but, as we have seen, it does not necessarily follow that where there is no smoke combustion is perfect.

In combustion the heat must be referred to the chemical union of the substances and the luminosity to the high temperature.

A jet of coal gas exhibits all the phenomena already described in regard to the flame of a candle; but, if the gas be previously mingled with air, or if air be forcibly mixed with or driven into the flame, no separation of carbon occurs, the hydrogen and carbon burn together, and the illuminating power almost disappears. Pure flame always emits a feeble light.

The perfect combustion of coal in a furnace can only be effected by a sufficient supply of oxygen, contained in the air, in a proper manner. To illustrate, suppose we have the fire lighted and steam raised in an ordinary wood-burning locomotive, and we throw into the furnace 500 pounds of bituminous coal; now, the engine being in motion, we shall observe some or all of the following results:

For a few minutes the flame in the fire-box will be darkened or nearly extinguished, the steam pressure falls, and a dense cloud of smoke and sparks come from the chimney.

After a little while the fire becomes bright, the amount of escaping smoke diminishes, and the steam pressure rises and attains its maximum.

There is now a strong local heat in the furnace, the temperature being higher than with wood. After a while the grate, perhaps, becomes clogged with clinker, and the surface of the fire, which at first had swelled as the coal was ignited, sinks in a crust of greater or less thickness.

If we allow the fire to burn down, and stop the engine, we shall find the tubes coated with soot and considerable fine coal in the smoke-box.

We find, on opening the fire-door soon after placing coal on the fire, that the smoke will be diminished both in quantity and intensity, although the steam pressure will fall, showing that the furnace is cooled by the admission of the air. Also the smoke which escapes from the chimney will deposit soot on any surface, as a piece of white paper, exposed to it. But the dense vapors formed in the fire-box will not, as they leave the coal, deposit any soot. So that, whatever may be the resemblance between the dark vapors, as distilled from the coal, and the smoke at the chimney top, they differ at least in one respect.

Again, the dark vapor in the furnace ignites and gives out flame as the air gains access to it, while the smoke, once formed, can not be burned by any practicable process—by heating, mixing with air, or otherwise. But while this vapor, which we shall now call gas, may be burned, it is not of itself combustible. If we fill a jar with this gas without admixture of air, a lighted candle, thrust into it, will be immediately extinguished. In ordinary gas-works, the same kind of gas is expelled from coal in cast-iron retorts heated to redness. It will burn only in contact with air. Now, while this gas burns regularly and silently when allowed to come into gradual mixture with the air—precisely as the same kind of gas burns in an ordinary illuminating burner—yet this same gas, *previously mixed* with about nine times its weight of air, will, when subsequently ignited, explode like gunpowder.

By "shutting off" steam from the engine, or closing the damper, the smoke is much increased; yet the only direct effect of the manipulation is the exclusion of air from the fire.

If a close jar be filled with the coal gas, and a jet of air be let into it, the jet of air can be ignited and will burn in the jar, exactly as a common gas-jet will burn in the air. The combustion of either is simply its chemical combination with the other.

PLATE 2.

While the gas is being expelled from the coal, the latter remains at a low heat—no particle of solid coal (or, rather, coke) can burn while gas is issuing from it. This, of course, refers to coal in its respective particles, as a *lump* of coal may be giving out gas in one place while all has been expelled from another, and the remaining coke already ignited.

This precedence in the burning of the gas is proved in making coke or charcoal. By admitting just sufficient air for the combustion of the gas in the raw fuel, this gas only is burned, and the coke or charcoal—pure carbon—is left behind. Were it not for this distinction in nature we could not manufacture either coke or charcoal except by distilling the gas from raw fuel in *close vessels*, and by the consumption of a considerable quantity of additional fuel in producing a high external heat.

As coke does not produce smoke in burning, the smoke issuing rom our experimental engine must be generated from the gas in combinations, which it forms before reaching the chimney. The duration of smoke, therefore, measures the time during which the gas is distilling from the coal.

If we examine closely we shall find that flame never enters a tube of ordinary size (1¼ to 3 inches) for more than a few inches from its mouth.

No matter how near the tubes are placed to the surface of the fire, the flame is extinguished immediately on entering them. Flame, as has been shown before, is superficial, and is *stripped*, so to speak, of its air on entering a small tube; and hence the core, or central body of gas (which, by its bulk, gave diameter to the flame), passes through unconsumed. No flame whatever can pass *through* a tube. An unignited compound, known as carbonic oxide, may pass through, and, having a low igniting temperature, may afterward take fire, and burn on coming to the air at the chimney-top. (The blue flame attending the conversion of carbonic oxide into

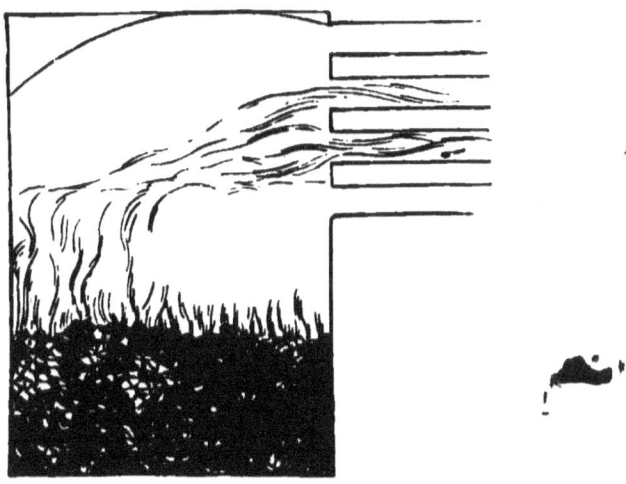

PLATE 3.

carbonic acid is not, however, the flame we have to deal with in the furnace.) Hence, all the combustible matter contained in the flame at the moment of its extinction is lost, as it can only impart that full heat by its complete combustion.

The intensity of heat from coal or any other fuel burned in air is a fixed and inevitable degree for each kind, and depends on the nature of its constituents.

By the perfect combustion of seasoned wood (which, however, contains about twenty per cent of water), the resulting heat is 2,867° Fahr.; bituminous coal, of average composition, 4,082°; anthracite coal, 4,170°; dry coke, or charcoal, 4,352°.

As long as water is in actual contact with the furnace sheets there is hardly any heat which can burn out the metal. . Water may be boiled over a fierce fire, when contained in a vessel the bottom of which is made of card paper, or it may be boiled in an egg-shell, and in neither case will the material of the vessel be injured.

In our experimental locomotive, with water spaces of, perhaps, not much more than two inches in width, we will find, by inserting a gauge-cock opposite the level of the surface of the coal, that water, if it issues at all, will do so intermittently when there is a very hot fire. This will account for much of the difficulty of burning out furnaces.

The effect of clinker and crust on the surface of the fire is to prevent the passage of air. The clinker is mainly sand or clay, vitrified or reduced by heat, and sometimes mixed with mineral oxides, which by their fusion still further increase its amount and strength. The quantity of clinker depends entirely upon the cleanliness of the coal. The only specific against its effects is to keep it well broken and cleared out. It is well to note, also, that the volatilization into gas abstracts considerable heat from the burning coal—the steam generally falling on the application of fresh coal to the fire.

The combustion of coal seems to be yet, to a great extent, a mystery even with those who profess to understand it, as appears from the contradictory plans proposed to effect complete combustion.

Chemistry, however, in a few simple principles, unfolds the whole philosophy of combustion.

PLATE 4.

Coal is a compound substance, and may be decomposed by heat into several distinct elements. As far as regards combustion we have to deal principally with two of these only, viz.: Carbon in the form of *coke*, and hydrogen, generally known as "*gas.*" These two elements practically contain the full heating properties of the coal. If we do not obstruct the processes is which they enter into combustion, they will naturally be completely burned, and consequently will impart their full measure of heat and will make no smoke.

Now, what is the operation of "burning"? We say that when coal is thrown upon a fire it begins to "burn." But *before* any burning can possibly commence, the coal must suffer the preparatory process of decomposition. Its constituent elements must be separated, and then a regular order of precedence obtains in their combustion. The burning, which then takes place, is this: the gas, which, having been distilled, burns first, does so merely by its chemical union with the invisible oxygen of the air, forming water (or steam). The gas having been burned, the coke takes its turn, and burns in exactly the same manner by combination with air, forming carbonic acid.

The combustion of the elements of coal in air is a mutual process, as is proved by the fact that neither of these by itself can possibly be burned. Flame itself is only the continuous explosion of successively combining atoms of air and gas, and which, had they been mixed previously in any considerable quantity, would detonate in the furnace like gunpowder, or with the force of a steam-boiler explosion. It is simply because this very mixture requires *time* for its accomplishment that the explosion can only go on by successive atoms, *thus forming continuous flame*. This influence of time must be kept distinctly in view, as it also determines, in a vital manner, the question of combustion of coal in steam-boiler furnaces. It is only in the condition of flame, in the natural process of combustion, that the gas and air can develop their useful heating power.

Now, if the gas does not have sufficient time to enter into complete mixture, atom by atom, with the air while *both* are within the range of an igniting temperature, they will produce *smoke*.

Coal is practically a solid compound of carbon and hydrogen. The carbon, so long as it remains as such, is always solid and visible; the hydrogen, when driven from the coal by heat, carries with it a portion of carbon, the gaseous compound being known as carbureted hydrogen. A ton of 2,000 pounds of average bituminous coal contains about 1,600 pounds, or 80 per cent of carbon; 100 pounds, or 5 per cent of hydrogen; and 300 pounds, or 15 per cent of oxygen, nitrogen, sulphur, sand, and ashes. But if this coal be *coked*, the 100 pounds of hydrogen driven off by heat will carry about 300 pounds of carbon in combination with it, making 400 pounds, or nearly 10,000 cubic feet of carbureted hydrogen gas. Thus, but 1,300 pounds of carbon (65 per cent of the original coal) will be left, and, with the earthy matter,—ashes, sulphur, etc., retained with it—the coke will weigh but about 1,350 or 1,400 pounds—67½ to 70 per cent of the original coal.

Thus, for every 2,000 pounds of coal we have about 1300 pounds of solid carbon and 400 pounds of carbureted hydrogen to be burned, the remaining 300 pounds being waste—partly gaseous and partly solid.

As we have seen, the combustion of this matter is simply its chemical combination with the oxygen of the air, as the nitrogen merely passes unchanged through the fire, effecting nothing but an abstraction of heat. The carbon and carbureted hydrogen in burning will each combine with oxygen in fixed proportions only, the proportions being known as chemical equivalents.

What we have to do, then, is *to permit, and not to obstruct, the access of air* to the coal, this access of air being permitted under such circumstances as shall *favor* its complete combination.

The only proportions in which carbon and hydrogen combine with air in combustion are these:

For every pound of carbon (pure coke) twelve pounds, equal to 159½ cubic feet, of air, are required to supply the necessary oxygen to combine intimately with it.

For every pound of hydrogen, thirty-six pounds, equal to 478½ cubic feet, of air, are required for similar combination. Thus for every pound of carbureted hydrogen gas, being one-fourth pound of hydrogen, and three-fourths pound of carbon, eighteen pounds, equal to 239¼ cubic feet, of air are required to supply the necessary oxygen.

Now, for every 2,000 pounds of coal burned, the 400 pounds of carbureted hydrogen—the "gas"—requires 95,700 cubic feet of air, at ordinary temperature, and the 1,300 pounds of solid carbon require 207,350 cubic feet of air. Practically, the "gas" from a ton of ordinary bituminous coal requires 100,000 cubic feet of air for its combustion, while the remaining coke requires 200,000 feet. Thus the gaseous matter of coal requires one-half as much air as is taken up by the solid coke. Where these combinations are completed the combustion is perfect. A given quantity of gas, completely burned, can not produce smoke, since smoke contains a quantity of unburnt matter, and is in itself a proof of incomplete combustion. The products of perfect combustion are invisible, being, for carbon and oxygen, carbonic acid; and for hydrogen and oxygen, invisible steam, which condenses into water.

102 GUIDE POSTS ON THE ENGINEER'S JOURNEY.

A coal-burning furnace is simply a vessel in which given elements are to be chemically combined in definite proportions. So far as the structure of the furnace may aid these combinations, by giving room, direction, and time for their completion, so far will the furnace be efficient in burning coal completely, which is the same as burning it economically.

Among the most essential provisions in the arrangement of the furnace are those which operate to induce or solicit the mingling of the coal-gas and air. Both these are to be burned, neither can burn without the other, they can only burn together. Thus, wherever the air is to be admitted, whether through the grates or through the openings in the door, the admission spaces should be small and numerous, so as to present as many surfaces of contact as possible. By diminishing the size and increasing the number of air spaces, we increase the surface of contact of the gas and air, just as, by diminishing the size and increasing the number of boiler tubes, we increase the surface of water exposed to heat. This increase of combining surfaces within a given bulk of gas and air, hastens the combination and hence the combustion of both.

The length of the flame proves how much time is required to complete the mixture of the combustible gas and air. Under a stationary boiler the flame has extended thirty feet from the grate. Flame, as before said, is entirely superficial, inclosing a core or central body of gas, waiting its turn to come into combination. This gas, having a progressive motion in the flues, or an ascensional power when discharged in the open air, must follow the draught, while by the continual combustion of its external atoms its thickness is being steadily reduced, until, if not violently extinguished, it finally burns out to a point and the flame terminates. Thus, if flame extends for thirty feet, the combining gas and air are mixing for the whole distance, and may not, even be fully mixed when the flame ceases,—what remains passing off unconsumed.

The want of time is experienced with the gaseous portions of the fuel only, as these, on the moment they are distilled, are on their flight to the tubes, giving but a fraction of a second for mixture and consequent combustion in a locomotive boiler. The solid portion remains quietly on the grate and takes its own time for combustion.

It may, indeed, happen, as is often the case, that even where the gases and air are detained in contact for a sufficient time for mixture, other arrangements of the boiler prevent their entering into perfect combustion. The process of the combustion of the carbureted hydrogen *must* be completed *before* it reaches the tubes. If, in a locomotive boiler, the fire-box does not afford room, it must be obtained by a combustion chamber in the barrel of the boiler, and even there we must have the requisite mixture by deflecting the currents of gas and air into as intimate contact as possible. We can not control the volume of the gas after the coal is on the grate, but we can divide the air into numberless thin streams, and by means of fire-brick bridges, or plates, etc., we can *deflect* the gas into mixture with the streams of air. These arrangements apply to gaseous fuels mainly, and not to coke, nor to the same extent to anthracite coal.

It is only by supplying and mixing the full equivalent of air that the combustible elements of the coal will be so combined as to give off their full heating power, and without the production of soot and its consequent smoke.

The best chemical authorities tell us that, in the most careful laboratory practice, a considerable time, often a whole day, is necessary for the complete mixture of certain gases; while those now under consideration can only mix after many minutes, and even then with an excess of oxygen, or, what is the same, an excess of air. That is, *more* air than the combining equivalent must be present in order that the combining equivalent may be taken up—although the excess of air, above this equivalent, is not burned but passes through uncombined.

Prof. Daniell, of England, says: "Any method of insuring the complete combustion of fuel, consisting partly of the volatile hydro-carbons (compounds of carbon and hydrogen), must be founded upon the principle of producing an intimate mixture with them of atmospheric air in *excess*, in that part of the furnace to which they naturally rise."

Practically in a furnace, nearly twice the air must be present that is actually required for the combustion of the gaseous matter.

As coal, in burning, does not give off its gas uniformly and continuously, but principally soon after being thrown upon the fire, we must have such control over the admission of air as will enable us to admit the right quantity according to the variable conditions within the furnace.

In continuous flame, the successively combining atoms of gas and air are ignited by contact, the process being described as a "self-generating succession," so long as both elements are supplied. The heat under which the gas itself distils will always ignite it, if the due admixture of air is immediately obtained. If the access or mixture of air is delayed until the gas has risen beyond the reach of an igniting temperature, it will then pass away unburned. And it is by far the more rational plan to effect the immediate admixture with air while the gas is ready to burn, than, neglecting this mixture, to endeavor to recover lost time by *heating* the gas afterwards, when it may be supposed to have fortuitously taken up its equivalent of air.

It is evident that to bring the coal, in both its solid and gaseous elements, into intimate mixture with air, and to ignite the compound, are all that human means can accomplish—nature only, in her own processes, effecting the rest. The distillation of gas, when fresh coal is supplied, goes on near the surface of the fire; the gas naturally burns above the surface, and the air necessary for its combustion must be admitted, therefore, above the surface. The necessity is this: Whatever amount of air passes through the grate will chemically combine (excepting a certain excess, which must always be present), with the solid part of the coal—the coke. The product of this combustion is carbonic acid, or carbonic oxide, and, although at a great heat, the oxygen which has entered into these compounds can serve no further combustible purpose. This carbonic acid is the same as that which, settling by its weight into the bottoms of wells or pits, *extinguishes* the flame of a candle. The fumes from coke or charcoal are carbonic acid only, and are as fatal to the existence of flame as to human life. The question is often asked, "Why not provide at once for the admission, through the grate, of sufficient air both for the coke and the gas?" This would be an impossibility, for whatever the quantity of air admitted through the grate, it will expend itself on the *coke* only—at least until holes are burned through the fire, when the control of the air is at once lost, and great waste of fuel ensues.

The admission of air above the fire must be in the greatest practical number of small jets, since gas and air mix only gradually, excepting by division and inducement. Air, *in bulk*, mixes only superficially with gas, and by abstracting heat cools the furnace. The air-holes should be placed as near as practicable to where the gases rise, since, after they are disengaged from the coal, it is necessary to *commence* their combustion at the earliest moment. Gases, to be thoroughly burned in the furnace, must be intercepted by air at the *start*, else the combination, which is at best gradual, will not be *completed* in season, as what remains uncombined on reaching the *tubes* is lost.

The temperature of the air entering the furnace has been a subject of controversy. No boiler furnace, however, was ever built in which hot air was proved to increase the intensity of combustion. The idea of hot air in this case has always been traced to the *seemingly* similar application in the hot-air blast furnace. There, however, the case is totally different. The iron lies among the coal, and the entering cold air, *before it has combined with the carbon*, chills the iron, thus rendering more coal necessary than if the iron had been heated. If, however, the air could combine with the coal *before it reached the iron*, its temperature (that of the air) would be quite a matter of indifference. In the boiler furnaces this combination takes place, or should do so, before the air reaches the plates, and the resulting heat is that generated directly within the air and coal, and can not be derived from any outside source.

There are certain practical objections to heating air. First, for every 480° of added heat, its bulk is enlarged by the amount of its original volume; so that at 3000°, the heat of the interior of the furnace, it has six times its original volume. It is consequently more unmanageable, and as its contained oxygen retains the same weight, its mixture with the gas becomes more difficult, while when mixed it can only do the same work as before. It would be much better to condense the air than to expand it. Next, if heated by passing through flame or over burning coal, the air will be robbed of a greater or less part of its vital oxygen. This is a positive loss.

The coldest air, if thoroughly mixed on its entrance with the coal or gases, will never cool, but will always sustain or increase the heat of the furnace. Air in bulk only can do any harm, and this is objectionable from the obstruction which it forms to combustion as well as from its abstraction of heat from the furnace. Thus, it is seen, the air, divided into thin streams, should be taken from the outside of the boiler directly into the furnace.

All that now remains to be provided for are *space* and such arrangements as shall give *direction* and *time* for combustion. The great volume of gas rising from the coal can not, even with the most ample admission and division of air, completely combine with it, except with prolonged opportunity and compulsory interfluence. By forcing both into convolution, in a circuit of several feet, we can effect their final union without injurious obstruction to the draught.

Finally, if all the combustible elements of the coal are completely burned, no smoke can be formed, and the evaporation effected will be all of which the coal is inherently capable—from eight to ten pounds of water for each pound of coal burned, according to its chemical constitution.

In flue or externally fired boilers the run of the gases is much longer, and thus *time* is given for a thorough admixture of air, if it be properly supplied; and in externally fired boilers, facilities exist for hanging bridges, etc., for the purpose of deflecting the gases, and thus affording a more thorough mixture of the air with them.

But it may as well be said that no matter how complete may be the arrangements of furnaces, grates, flues, bridges, etc., as well as air supply, a careless or ignorant fireman will waste coal, the waste in some cases amounting to more than thirty per cent.

To sum up, the quantity of air required for combustion varies with the composition of the fuel; but it is sufficient for practical accuracy to say that each pound of fuel requires twelve pounds of air. The volume of air, of course, will depend upon its temperature.

Now the quantity of heat which can be developed by one pound of pure carbon is sufficient to boil fifteen pounds of water from and at 212° Fahrenheit, if none of the heat were lost; but there are many reasons for non-attainment of this result in practice, and the following are some of them:

First. Differences in the chemical constituents of the coal.
Second. Impurities present in the coal.
Third. Losses by conduction and radiation in the furnaces, flues, and metal of the boiler.
Fourth. Imperfect and incomplete combustion.
Fifth. Loss of heat carried off by the chimney, partly utilized in causing the draught.
Sixth. Improper management.

There is a great difference between anthracite and bituminous coals as fuel. Anthracite burns completely with a thin fire, by admitting an excess of air through it and above it; but bituminous coal absolutely requires for its perfect combustion a high temperature and plenty of room for the products of combustion, before coming into contact with the iron of the boiler, with a proper supply of air above the fuel; and any deviation from these conditions produces smoke and loss of heat.

With hard coal, too great a draught wastes a little heat in the chimney; but with soft coal, too great a draught may be as bad in its effects as not enough. The great secret in smoke-prevention is to have a hot fire with plenty of *room* and *time* to let all the gas burn before getting lower in temperature than a red heat (800° Fahr.), and to fire in small quantities over a part of the grate at a time.

GUIDE POSTS ON THE ENGINEER'S JOURNEY.   105

In a chimney where the draught is produced by the excess of weight of the outside air over that of the hot gas in the chimney, the greatest quantity of gas by weight will pass up the chimney when its temperature is about 625° greater than that of the outside air. But it is a well-known fact that natural draught is not so economical as a forced draught, because a certain amount of heat is wasted in producing this draught—about twenty-five per cent—and the cost of a forced draught to burn the same amount of coal in the same time is not half so great.

The power of boilers is much increased by a forced draught, the comparative efficiency being as follows:

$$\begin{aligned}\text{With Natural Draught} &= 1\\ \text{`` Jet} &= 1.25\\ \text{`` Blast} &= 1.6\end{aligned}$$

TOTAL HEAT EVOLVED BY COMBUSTIBLES, AND THEIR EQUIVALENT EVAPORATIVE POWER, WITH THE WEIGHT OF OXYGEN AND QUANTITY OF AIR CHEMICALLY CONSUMED.

COMBUSTIBLE.	Weight of oxygen consumed per lb. of combustible.	Quantity of air consumed per pound of combustible.		Total heat of combustion of 1 lb. of combustible.	Equivalent evaporative power of 1 pound of combustible, under the atmosphere at 212° Fahr.
1 lb. weight.	Lb.	Lb.	Cub. ft. at 60° Fahr.	Units.	Lbs. from and at 212° Fahr.
Hydrogen........................	8.0	34.8	457	62,032	64.2
Carbon, making Carbonic Oxide......	1.33	5.8	76	4,452	4.61
Carbon, making Carbonic Acid	2.66	11.6	152	14,500	15.0
Carbonic Oxide......................	0.57	2.48	33	4,325	4.48
Light Carbureted Hydrogen	4.0	17.4	229	23,513	24.34
Bi-Carbur'd Hydrogen, or Olefiant Gas	3.43	15.0	196	21,343	22.00
Sulphur...........................	1.00	4.35	57	4,032	4.17
Coal of average Composition.........	2.46	10.7	140	14,133	14.62
Coke, desiccated....................	2.50	10.9	143	13,550	14.02
Wood..............................	1.40	6.1	80	7,792	8.07
Peat...............................	1.75	7.6	100	9,951	10.30
Lignite.............................	2.03	8.85	116	11,678	12.10
Asphalt............................	2.73	11.87	156	16,655	17.24
Straw, 15% per cent moisture........	.98	4.26	56	5,196	5.56
Petroleum.........................	4.12	17.93	235	27,531	28.50

INCRUSTATION.

Water is rarely found pure, *i. e.*, not holding foreign substances in solution. Consequently, when water is poured into a vessel placed over a fire, and is boiled until it all escapes as steam, we shall find on the bottom and sides of the vessel a coating or deposit, which, if *hard*, we call incrustation or scale; but if it is in the form of a powder, we call it sediment. Sometimes, also, the water holds mineral matters in a state of mechanical suspension.

Fatty matters, introduced with the feed water, sometimes cause a very troublesome incrustation, or a soapy deposit.

The substances more generally found in solution in water are the carbonates of lime and magnesia, sulphate of lime, chloride of magnesia, with traces of silica, alumina, and iron.

Incrustation and sediment in a boiler cause waste of fuel, are a source of danger, and are oftentimes expensive to remove.

Water can be freed from substances held in mechanical suspension by filtration, but it is often both difficult and expensive to remove those chemically combined with it.

Heat is the only agent by which all the matter held in solution can be removed from the water.

From ten to forty grains of mineral matter per gallon of 231 cubic inches are held in solution by the water of rivers, streams, and lakes; well and mine waters contain more.

Since water contains all the incrustating ingredients found in boilers, having been the means of conveying them there, it will be well to treat of its composition and properties.

Water is composed of two gases, hydrogen and oxygen, in proportion, by weight, of 11.1 parts of hydrogen and 88.9 parts of oxygen; or, in chemical combination, of two atoms of hydrogen and one of oxygen.

Water, when pure and in small quantities, is transparent, colorless, tasteless, odorless, and is a bad conductor of heat and electricity. Under a pressure of 30,000 pounds to the square inch, fourteen volumes of it may be condensed into thirteen volumes, so that it is slightly elastic. It is eight hundred and fifteen times heavier than atmospheric air, an imperial gallon (277.274 cubic inches) weighing, at 62° Fahrenheit, and under a barometric pressure of 29.92 inches, 70,000 grains, or ten pounds avoirdupois; but as it is the standard to which the gravities of solids and fluids are referred, its specific weight is called 1.

Water contracts and becomes denser in cooling until it reaches 39.2° Fahrenheit, when it has reached its greatest density. Below this point it expands, and at 32° Fahrenheit it becomes solid, or freezes, and in the act of freezing expands considerably. Owing to the expansion, ice is lighter than water, it having a specific gravity of 0.916. The bursting of water-pipes is often due to the expansion of the water in them while freezing.

Water is expanded by heat between 40° and 212° Fahrenheit, boiling in metallic vessels at the latter temperatures when the barometric pressure is 29.92 inches.

The boiling point of water varies according to the pressure. As at a pressure of 29.92 inches it boils at 212° Fahrenheit, so at a pressure of only 27.74 inches it boils at 208° Fahrenheit.

The solvent properties of water far exceed those of any other known liquid. A very large proportion of the different salts are more or less soluble in it, the solubility increasing generally as the temperature rises, so that a hot saturated solution deposits crystals on cooling.

GUIDE POSTS ON THE ENGINEER'S JOURNEY. 107

There are a few exceptions to this rule, one of which is common salt, the solubility of which is about the same at all temperatures; the hydrate of lime (slaked lime) is more soluble in cold than in hot water; sulphate of lime is also less soluble in hot than in cold water, and wholly insoluble at 302° Fahrenheit, or between 284 and 302° Fahrenheit. Water also dissolves gases, but a boiling temperature expels all the gas if it be not very soluble. The solvent properties of water are still further increased when heated in a strong vessel under pressure.

Water containing carbonate of lime, held in solution by free carbonic acid, boils steadily, and is not liable to cause foaming. As the water boils the carbonic acid gradually escapes, the carbonate of lime then being deposited in the insoluble, and frequently in the crystalline state. The more slowly it is deposited the more crystalline it will be, sometimes becoming hard like a rock and requiring to be chipped off.

The incrustation generally contains some sulphate of lime, it being less soluble in hot than in cold water. This incrustation gradually eats away a portion of the metal, to which it becomes so tightly fastened that when it is chipped off there will always be found a layer of oxide of iron on that side of the incrustation which was in contact with the metal. The corrosive effects produced by water holding salts in solution take place on the iron of the boiler chiefly, but *well water* containing a quantity of soluble salts acts upon the brass and copper to such an extent that boilers are almost covered in front with incrustations of the salts; of course greatly injuring and in some cases rapidly destroying them.

Incrustation is commonly stated to be a bad conductor of heat, and that any great thickness of it on the plates of a boiler causes a largely increased expenditure of fuel. It is not clearly determined yet whether the increased expenditure of fuel is quite so great as has been claimed, though it is undoubtedly a source of waste.

There are many compounds for preventing incrustation, some are very costly, and none of them are alike applicable to all cases. Some of them increase corrosion of the boilers while they prevent incrustation, and in the employment of others practical difficulties arise which render their use inconvenient and, at times, even dangerous.

Dr. Kossman states that zinc introduced into steam boilers to prevent incrustation proves very useful in case of selenitic waters, but as against the carbonates of lime, magnesia, and iron, it is of little value, the zinc being soon rendered brittle and porous, and in a short time reduced to powder.

Zinc has also been used in marine boilers with varying degrees of success. The most important effect, however, obtained from its use, is the protection of the plates, etc., from the hydrochloric acid evolved from the chloride of magnesium of the sea-water, as stated by Dr. Gideon E. Moore, who analyzed specimens of zinc and scale from the boilers of the steamer "Rosedale," plying on the waters of Long Island Sound.

For the removal of scale already deposited in boilers, as well as preventing its formation without injury to the boiler, the Basic Tri-Sodium-Phosphate, manufactured by the Keystone Chemical Company, of Philadelphia, has attained a remarkable reputation in these respects. We can say with truth, the most careful analysis by eminent chemists has shown that it is incapable of injuring the iron of a boiler, and use has emphatically confirmed this statement.

Where boilers are already coated with hard scale, its use gradually converts the stony incrustations of the carbonates of lime and magnesia, and even anhydrous sulphate of lime, into pulverulent and flocculent phosphates. It also separates the foreign matter held in solution by the water as a light flocculent precipitate, which will not bake into a crystalline scale, but is easily removed by blowing off, thus preventing the formation of incrustations; and, further more, it neutralizes acids contained in the water, thus rendering them innocuous.

It is needless to say that its use promotes economy of fuel.

GUIDE POSTS ON THE ENGINEER'S JOURNEY.

In the following table is shown the varying amounts of mineral and organic matter held in solution by the waters of different localities:

ANALYSIS OF WATER BY PROF. C. F. CHANDLER, COLUMBIA COLLEGE.

TABLE SHOWING WEIGHT OF IMPURITIES, IN GRAINS, IN 1 U. S. GALLON, 231 CUBIC INCHES.

No.	Source.	Corroding matter.	Incrusting matter.	Organic matter.	Total solids.
1	Syracuse, Onondaga Creek	3.44	22.58	0.34	26.36
2	" hydrant	0.38	27.55	trace.	27.93
3	Memphis	0.91	21.68	0.18	22.77
4	Jordan	1.71	11.47	0.06	13.24
5	Port Byron	1.08	7.17	1.28	9.53
6	Savannah	1.35	17.63	1.52	20.50
7	Clyde, spring	0.77	14.64	2.16	17.58
8	" River	2.10	14.30	1.88	18.28
9	Lyons	1.03	11.07	1.00	13.10
10	Newark	1.17	18.73	2.16	22.07
11	Palmyra	1.43	33.39	1.46	36.28
12	Macedon Swamp	0.71	10.53	0.80	12.04
13	Fairport	3.19	15.06	1.14	19.39
14	Rochester, North Street well	7.31	33.26	1.60	42.17
15	" Genesee River	1.18	10.85	1.64	13.67
16	" Canal, round-house	1.11	8.80	1.24	11.15
	Incrustation preventives.				
17	Warner's 8.26	3.72	11.28	0.37	23.63

These places are on the line of the New York Central R.R.

The corroding matters are: the chlorides of potassium, sodium and manganese, and the sulphates of potassa and soda. The incrusting matters are: the sulphate of lime, the carbonates of lime and magnesia, the oxide of iron, and silica. The incrustation preventives are: the carbonates of soda and potassa

Six specimens of incrustations analyzed by Prof. Chandler give as an average composition:

Sulphate of Lime	56.49
Carbonate of Lime	18.11
Basic Carbonate of Magnesia	19.77
Oxide of Iron and Alumina	0.69
Silica	3.81
Organic Matter	undetermined
Water	1.62
	100.00

These analyses show that the incrustations consisted chiefly of the carbonates of lime and magnesia, and the sulphate of lime.

The two carbonates are insoluble in pure water, and owe their presence in the waters of springs and rivers to free carbonic acid, which forms with them soluble bicarbonates.

Boiling such waters expels this carbonic acid, and the carbonates of lime and magnesia separate in the form of insoluble powders. Now, if this boiling takes place in an ordinary steam boiler, these carbonates are set free as impalpable powder and buoyed to the surface of the water, where they remain for some time, gradually being carried to the comparatively cooler and quieter parts of the boiler, where they quietly and gradually settle, forming a sludge. Were some suitable means provided, they could be collected at the surface and blown out at intervals, without much loss of efficiency to the boiler.

Various alkaline substances, by appropriating the carbonic acid, cause the precipitation of the insoluble carbonates. Potash, soda and ammonia, as well as their carbonates, produce this effect. So that the carbonates may be removed without decomposition by simply depriving them, of their solvent, the carbonic acid, and this can be done in tanks, and the purified water fed to the boiler; but the process requires care.

The sulphate of lime is soluble in water, one gallon of water being capable of holding 150 grains in solution; but the solubility of the sulphate of lime in water is modified by the presence of other substances. The chlorides of calcium and magnesium, and even a high temperature, diminish, while the chlorides of sodium and ammonium, and various organic substances, increase its solubility. Above 212° Fahr. the solubility rapidly diminishes as the temperature increases, and at 302° Fahr., equivalent to a pressure of 70 pounds per square inch, it may be said to be totally insoluble, and, in fact, at a temperature of 250° Fahr., equivalent to a pressure of 30 pounds, so much of the sulphate of lime is rendered insoluble that the quantity remaining in solution is not practically objectionable.

Sulphate of lime forms a hard crystalline scale in the absence of carbonates. When the carbonates of lime and magnesia are present, the deposits vary from a loose powder to a hard crystalline formation, according to the proportions of the three substances.

In practice, sulphate of lime can only be removed from water by undergoing decomposition. For instance: by carbonate of soda, which forms carbonate of lime, which is deposited as powder and sulphate of soda, which remains in solution. But when soda is used it is advisable to have some surface-blow apparatus fitted to the boiler to avoid trouble from "foaming."

Incrustations are injurious to boilers because—

1st. They are poor conductors of heat, and so cause loss of heat, and, as a consequence, waste of fuel. The estimated waste is from fifteen up to forty per centum of the fuel used, dependent upon the thickness of the scale.

2d. They cause overheating of the boiler plates, which is sure, sooner or later, to cause a burning out of the metal that may result in an explosion of the boiler.

3d. The corrosion of the metal occurs most rapidly in those parts of the boiler upon which the deposits are more liable to accumulate.

The following table shows the amount of foreign matter held in solution in the waters of different parts of the New England States:

ANALYSIS OF WATER BY S. DANA HAYES.

TABLE SHOWING WEIGHT OF IMPURITIES, IN GRAINS, IN ONE U. S. GALLON, 231 CUBIC INCHES.

No.	Source.	Mineral matter.	Organic matter.	Total Solids.
	MAINE.			
1	Pure spring, near Auburn	0.85	0.13	0.98
2	Spring on Cape Elizabeth	7.40	2.21	9.61
3	Wells in Portland (average of four)	13.35	5.13	18.48
	NEW HAMPSHIRE.			
4	Merrimac River, at Manchester (drainage)	2.96	2.60	5.56
5	Merrimac River, at Lowell, Mass.	1.80	0.11	1.91
6	Massabeesic Lake, near Manchester	1.16	1.66	2.82
7	Hotel well, on Rye Beach	6.08	2.43	8.51
	VERMONT.			
8	Mineral Springs, near St. Albans (average of seven)	15.24	1.25	16.49
9	Mineral Springs, at Guilford (chalybeate)	25.27	1.65	26.92
10	" " at Brunswick	77.79	2.33	80.12
11	" " at Danby	7.19	0.91	8.10
	MASSACHUSETTS.			
12	Cochituate, Boston, February, 1871	2.37	0.83	3.20
13	Mystic, Charlestown, February, 1871	3.96	1.72	5.68
14	Jamaica Pond, Roxbury, 1867	2.41	1.36	3.77
15	Connecticut River, at Holyoke	1.81	1.39	3.20
16	Saugus River, Lynn	3.12	2.40	5.52
17	Flax Pond, Lynn (drainage)	2.24	1.84	4.08
18	Horn Pond, Woburn	3.85	1.59	5.44
19	Locomotive supply, Taunton	4.37	2.03	6.40
20	Artesian well, Dedham	4.08	1.11	5.19
21	Wells in Woburn (average of four)	51.52	4.60	56.12
22	Wells in Lynn (average of six)	19.27	4.23	23.50
23	Old Artesian well, Boston (reopened 1871)	54.35	1.85	56.20
24	Well on Cape Cod	10.01	2.41	12.42
25	Brewery Spring, Boston	13.68	1.68	15.36

So far, incrustations due to the deposition of matter held in solution in the water only have been spoken of, as those are by far the most frequent. But it is known that a quantity of fatty matter in the feed-water of a boiler causes a deposit which is not wet by water, and that this may lead to the destruction of the boiler, inasmuch as the part of the boiler shell, under the incrustation, becomes more heated than other parts, and is apt to occasion rupture.

Several cases have come under the notice of the United States Steamboat Inspectors where oil, used to lubricate the cylinders, has been carried into the boilers with the feed-water, and settling upon the furnace crowns, has formed a thin, hard incrustation there, which resulted in burning the plates, followed by a crack, necessitating repairs.

There is also found a chemical combination of the fatty acids of the oil with the impalpable powder of the carbonates of lime and magnesia, forming an insoluble soap, which agglomerates, and finally sinks to the bottom of the boiler, and being plastic, closely conforms to the shape of the sheets, thoroughly excluding the water from them, and generally resulting in the formation of "bagging," or "pockets," which will soon burn and crack, and perhaps cause a dangerous rupture.

An instance of this kind was lately discovered by an Inspector of the American Steam Boiler Insurance Company. The incrustation had a soapy feel, was of a chocolate color, and varied from one-half inch to several inches in thickness, and the "bulging" or "bagging" of the plates was very pronounced.

There is no doubt that the use of apparatus for collecting the fine particles of minerals, as they are freed from solution by heat, pressure and chemical reactions in the boiler, while they are still near the surface of the water, and then periodically blowing them out, will certainly greatly reduce the amount of incrustation, conducing to economy, and adding to the life and efficiency of the boiler.

Incrustation is a source of danger also, as it clogs the feed and blow-pipes, and also the water-legs in fire-box boilers. And again, when a boiler is much incrusted, it is impossible to examine so thoroughly as to ascertain the true condition of the plates, joints, rivets, etc. Of course this incrustation must be removed by the use of chisels, picks, etc., which tends to injure and weaken the structure.

In the alkali regions of the United States it is very hard to keep boilers in a passable condition, as the water is strongly impregnated with mineral substances. Accounts are given of the use of kerosene oil for the purpose of removing scale, and apparently with success. But it is an article that must be used with great care and judgement, or the boiler will soon be ruined. Again as before mentioned, much of the deposit in a boiler lies in the bottom as a sludge, and should be removed from the boiler as such. But when fires are hauled at the end of a week's work, steam blown off, and the water blown out of the boiler, on examining the boiler within twenty-four hour, it will be found that the heat of the iron and brick work has baked that sludge into a hard scales very difficult to remove.

In view of what has been shown in regard to incrustation, the following is strongly recommended:

1. The use of the purest water that can be obtained.
2. Frequent use of a surface blow-cock.
3. That the boilers never be emptied until they are so cool that the deposits will not bake upon the metal.
4. Frequent examinations and cleanings.

When the feed-water produces much incrustation, it has been sometimes found that, by using rain water two or three times a week in the boilers in place of ordinary feed, a surprising effect is produced, as the scale already there loosens and drops off, and, in addition, very little scale forms thereafter.

It may be as well to show how an engineer can ascertain for himself what oreign matters are present in the water used in his boilers. For this purpose he only requires a few test tubes and a few small bottles of chemicals, with a test-tube holder and some litmus paper. Below is a list of what is required:

LIST OF CHEMICALS AND APPARATUS.

½ pint bottle of Soap Solution.
1 2-oz. " " Lime Water.
1 2-oz. " " Chloride of Barium.
1 2-oz. " " Chloride of Ammonium.
1 2-oz. " " Ferrocyanide of Potassium.
1 2-oz. " " Hydrochloric Acid.
1 2-oz. " " Nitric Acid.
1 2-oz. " " Tincture of Cochineal.
1 2-oz. " " Metallic Mercury.
1 2-oz. " " Carbonate Ammonia (crystals).

1 1-oz. bottle of Oxalic Acid (crystals).
1 " " " Phosphate of Soda (crystals).
Slips of Blue Litmus Paper.
Slips of Red Litmus Paper.
1 4-oz. flat bottom clear glass Bottle.
A Wooden Test-Tube Holder.
One small Spirit Lamp.
½ pint of Alcohol.
A Test-Tube Brush.
½ doz. Test Tubes.

These can be supplied by any chemist.

Take any clean bottle and fill it with the water you desire to test, and proceed as follows:

TO SEE WHETHER THE WATER IS HARD OR SOFT.

Take a clean test-tube and pour into it about three-quarters of an inch in depth of the soap solution; then pour into it three or four drops only of the water; if it becomes milky or curdy the water is hard.

TO SEE IF THE WATER IS ALKALINE OR ACID.

Dip into a test-tube half filled with the water, a strip of red litmus paper; if it does not turn blue the water is not alkaline. Now dip a strip of blue litmus paper into the water; if it does not turn red the water is not acid.

TO SEE IF THERE IS CARBONIC ACID.

Fill about three-quarters of an inch of water into a test tube, and then pour in just as much lime water; if there is carbonic acid the water will become milky, and on adding a little hydrochloric acid the water will become clear again.

TEST FOR SULPHATE OF LIME (Gypsum).

Fill in the water to the depth of one and one-half inches in a test-tube, and then add a little chloride of barium; if a white precipitate is formed, and it will not redissolve when you add a little nitric acid, sulphate of lime is present.

TEST FOR MAGNESIA.

Fill a test-tube about one-fourth or one-third full with the water; hold it with tube holder, and bring it to a boil over the spirit lamp; then add the point of a knife full of carbonate of ammonia, and a very little phosphate of soda; if magnesia is present it will form a white precipitate; but as it may not do so at once, it is best to set it one side for a few moments.

TEST FOR LEAD.

Fill a test-tube one-fourth full of the water, and add one or two drops only of tincture of cochineal. If there be only a trace of lead in the water, it will be colored *blue* instead of pink.

TEST FOR COPPER.

Add to some water in a test-tube a little filing dust of soft iron, and a few drops of chloride of ammonium; a blue colorization denotes the presence of copper.

TEST FOR IRON.

To some water in a test-tube add one drop of ferrocyanide of potassium; it will color it blue if iron be present.

TEST FOR SULPHUR COMBINATIONS.

Pour enough mercury into a small glass bottle with a flat bottom to cover it, then pour in water enough to fill it for a depth of half an inch or more, stopper the bottle and let it stand a few hours. If the mercury assumes a darker surface, and upon shaking separates into a dark powder, the water contains sulphur combinations.

Remember to rinse a test-tube out thoroughly before using with the water that you are about to test, and after making one test, rinse out the tube thoroughly in the water, using the tube-brush if necessary.

The soap solution can be prepared by putting some fine scrapings of white curd soap (from an apothecary) into a bottle and pouring alcohol upon it, then cork the bottle and set it one side, shaking it often for a few days until it is all dissolved, then add a little more soap, and if you find you have too much, add a little alcohol, so as to just dissolve it.

Lime water can be prepared by slaking a small lump of freshly burned lime with half its weight of water in a vegetable dish; then take some of the slaked lime and put it in a bottle with some cold distilled water (which can be obtained by condensing steam), and shaking it occasionally; then let the undissolved portion subside, draw off most of the clear liquid, and keep it tightly stoppered in a clean bottle.

NOTE.—Lime water shaken up with linseed oil in a bottle forms a yellowish, creamy substance, which is a very soothing and cooling application in case of severe burns and scalds.

So far mention only has been made of internal incrustations due to foreign matter held in solution by the feed-water; but there are also deposits forming incrustations in the tubes and flues.

Professor Hayes, in speaking of the deposits in tubes and flues, says:

"They are of two kinds, both of which are capable of corroding the iron rapidly, especially when the boilers are heated and in operation. The most common one consists of soot (nearly pure carbon) saturated with pyroligneous acid, and contains a large proportion of iron if the deposit is an old one, or very little if it has been recently formed. The other has a basis of soot and fine coal ashes (silicate of alumina), filled with sulphur acids, and containing more or less iron, the quantity depending on the age of the deposit. The pyroligneous deposits are always occasioned by want of judgment in kindling and managing the fires. The boilers being cold, the fires are generally started with wood; pyroligneous acid then distils over into the tubes, and collecting with the soot already there from the first kindling fires, forms the nucleus for the deposits, which soon become permanent and more dangerous every time wood is used in the furnace afterwards.

"The sulphur-acid deposits derive their acids from the coal used, but the basis material holding these acids is at first occasioned by cleaning or shaking the grates soon after adding fresh charges of coal. Fine ashes are thus driven into the flues at the opportune moment for them to become absorbents for the sulphur compounds distilling from the coals, and the corrosion of the iron follows rapidly after the formation of these deposits."

It is well to remark that the above-mentioned deposits orm a very hard incrustation, though of but little thickness generally, and that they are very bad conductors of heat; therefore their removal is a necessity.

CORROSION.

Corrosion is the strongest destructive force to which a boiler is subjected, and is of two kinds, external and internal.

Internal corrosion can be divided into three classes, known as: uniform corrosion, or wasting; pitting, or honey-combing, and grooving; so named from the appearances they present.

Uniform corrosion is that species of wasting of the plates, tubes, etc., where the water corrodes them, in a more or less even manner, in patches of large extent, and where there is usually no well-defined line of demarcation between the sound and corrodedp arts. It is like ordinary rusting in its character and effects, but is seldom so uniform. This is easily detected, and, even when covered by a thick coating of incrustation, on emptying a boiler, it is shown by red streaks where the scale is cracked, or "bleeding," as it is sometimes called; but when detected, owing to its uniform appearance, the depth to which it has penetrated can only be determined by drilling holes through the plate and measuring the thickness.

Corrosion is apparently very capricious in its action. Two boilers, made exactly alike, of the same iron, fed with the same water, and subjected to the same amount of work, will be differently affected—one may be attacked in the bottom, the other at or above the water-line. Doubtless the differences in the qualities of the plates have much to do with this. From experiments made by Drs. Whelpley and Storer some years since, it was found that, in specimens cut from the same plate of iron, the best transmitting qualities varied by from ten to forty per cent, showing decidedly that the structure and qualities of a plate of iron are not homogeneous.

Again, sometimes boilers are attacked on the shell, at others they will suffer principally in the tubes, and at others the rivets and seams suffer most; and sometimes the stays waste more rapidly than any other portion. This action, so erratic, of the corrosive agents must be ascribed to their gravity, to their concentration in certain parts, to the circulation of the water, to the nature of the iron, and to other more obscure causes.

In many boilers, supplied with water containing sulphate of lime, when scale is detached, a black coating of oxide of iron is found adhering to the scale, and as often as the scale is reformed and detached, it brings with it a fresh film of the oxide.

Another peculiarity may be noticed—that whereas, in most cases, the corroded iron is readily removed, yet cases often occur where the corroded iron adheres most tenaciously to the sound iron beneath it, so that considerable force is required to remove it, and the presence of corrosion may not be even suspected until the pick or hammer is forcibly applied.

Pitting and honey-combing are well marked by the sharply defined edges they present. The term honey-combing is generally used to describe the appearance of plates indented by very small holes close together. Pitting is confluent honey-combing, and is found in patches of various sizes, varying from an inch to twelve inches or more in diameter, and from one-sixteenth to one-fourth of an inch in depth, and of very irregular forms.

This form of corrosion has been thoroughly investigated, but no satisfactory reason has been rendered for it. Galvanic action was once supposed to be the prime cause of pitting, but that theory has been dropped, and it is now generally conceded that pitting is due to simple chemical action.

It is well known that the concentrated acids of the water will attack the most susceptible portions of the plates; and if the acids are volatile, or the liquid acids are carried, by foaming and priming, into the steam space, the plates there also suffer. The wasting of plates round the holes is principally due to the injury to the material when punching the holes, which render the metal more susceptible to the action of the water.

The surest way to prevent internal corrosion is to abandon the use of water which produces that effect, but this is not always entirely practicable.

When the water is found to affect the plates only in particular places, as at the water-level, it is well to use thicker plates at such places, and to arrange them so that the seams do not come within the region attacked by the water. When there is no choice of feed-water, the acids may be neutralized and corrosion prevented by the use of some alkaline substance prior to the introduction of the feed-water into the boiler. This is best done by using soda ash, or carbonate of soda, which should be dissolved and enter the boiler with the feed-water. The quantity required will vary according to the strength and quantity of the acids in the water, and can be best ascertained by experiment.

Grooving, channeling, or furrowing, as it is variously termed, is due to mechanical action produced by unequal expansion and contraction. Where it is not aided by the corrosive action o the water, it may penetrate deeply into the plate without being more than one-sixteenth of an inch wide at the surface, and sometimes the grooving is so fine that it appears like a fracture. The introduction of the feed-water near the water-level in the boiler, instead of near the bottom, helps materially to prevent grooving.

Internal grooving is undoubtedly often introduced by careless or excessive caulking, which destroys the skin of the iron, and thus exposes a surface to the attack of the corrosive agents in the water.

External corrosion is a more subtle agent in the destruction of stationary boilers than any kind of internal corrosion, and this is because its presence is less suspected, and it is not easily to be detected on account of the difficulty of getting at the plates.

Improper setting in brick-work often causes corrosion, the part of the boiler shell exposed to the action of the probably impure lime having been badly eaten; but this can be prevented by using fire-clay, instead of common mortar, as a cement in such places. And external corrosion is also caused by exposure to the weather, leakage at joints, leakage of fittings, drippings from pipes, etc., moisture from wetting down ashes near the boiler, and moisture rising from the ground, etc. Cooling off the boiler too rapidly and filling it while it is still warm are productive of leaky seams, and hasten the destruction of a boiler.

The tube ends are a source of annoyance in some types of boilers, and this is especially apt to be the case with vertical boilers, as the upper ends are exposed to the action of the heated gases, and there being no water to prevent overheating, they are soon loosened, and commence to leak badly This gives rise to corrosion of the ends of the tubes and the upper head, which, in many cases, goes on with great rapidity. It is no unusual thing to find the upper tube-sheet of upright boilers eaten half way through, and nearly all the tubes leaking badly. This leakage is not so apparent from steam pressure as from water pressure. To the unpractised boiler attendant everything may appear all right, but when the boiler is filled to the top with water, and pressure is applied, there is generally trouble.

The lower ends of the tubes are also apt to give trouble, especially when the upright boilers are used for heating purposes, and the blow-off does not drain the boiler; for, during the summer months, when the boiler is idle, the interior of the shell and tubes, just at the surface of the water left in the boiler, are subjected to severe pitting. Sometimes the tubes of this class of boilers are completely riddled in a very few seasons, whereas, if properly cared for, they should last several years. Where the upper ends of tubes are loosened from the action of heat they may be made tight by the use of an expander if not too much corroded, but if the body of the tube is pitted also, the only remedy is a new tube.

BOILER MAKING.

The practical building of a boiler in a shop is somewhat after the following manner:

When a contract has been made to build a boiler according to specifications furnished, the first thing to be done is to order the necessary iron from the mill. Now, suppose the boiler is to be 60 inches in diameter, that is, the diameter inside the large course, and if the boiler courses are each to be made of one sheet only, as is much preferable, in order to have a length of sheet sufficient to make a ring of that diameter, we must add the thickness of the sheet to that diameter and multiply their sum by 3.1416; suppose the thickness of the sheet to be three-eighths of an inch, then their sum is 60 + ⅜ = 60⅜ inches, or 60.375 inches, and this gives us (60.375 × 3.1416) a length of 189.67 inches; but there is no provision for lap to make a joint, and as the usual provision for a double-riveted joint, which this should be, is 4½ inches, we must add this amount to the length already found, which will give us 189.67 + 4.5 = 194.17 inches, or, to allow for trimming, 194.5 inches. By referring to the drawing accompanying the specifications we shall find the width of the sheet between the centers of the girth seams of rivets, suppose this to be 72 inches; now as these girth seams are single riveted, the distance from the center of rivet hole to the edge of the lap will be 1⅛ inches, so we must add the double of this, or 2¼ inches, to 72 inches, to get the proper width of the sheet, and 72 + 2.25 = 74.25, and adding for trimming, we should have about 75 inches. So that we must order for this large course a sheet of 194.5 by 75 inches from the mill, specifying in the order the quality and tensile strength.

In calculating the length of the inner or smaller course we must deduct its thickness from the boiler diameter, and we shall have 60 − .375 = 59.625, and 59.625 × 3.1416 = 187.31 inches and adding for lap 4½ inches, we have 187.31 + 4.5 = 191.81 inches, and adding for trimming we shall have 192.5 inches; and supposing all the courses to have the same width, our order for the small course sheets would require plates of 192.5 by 75 inches.

The heads are usually ordered to have the diameter of the boiler, it being understood that a disk of 5 inches larger in diameter will be sent, thus giving a flange of 2½ inches. The thickness and quality of the head must be specified in the order.

When the iron arrives at the shop, it should be examined for scabs, pitts, cracks, etc., and then each sheet should be tested for blisters, and if the boiler is to be built under the supervision of the American Steam Boiler Insurance Co., the Inspection Department of that company, at 79 John Street, should be notified at once.

The best way to test a sheet for blisters is to support it horizontally at the four corners, and then cover its upper surface with a layer of fine dry sand, then by tapping the under surface all over with a light hammer, the sand on the upper surface will be put in motion by the vibrations wherever the iron is solid, and the position of a blister will be indicated by the sand remaining undisturbed by the blows of the hammer.

Another method is to rule both sides of the sheet with a slate pencil into two-inch squares, and then, after suspending the sheet, to tap in the center of each square on both sides of the sheet with a light hammer, the difference in the sound denoting whether the hammer strikes solid iron or not.

Meanwhile a strip of iron taken rom one of the sheets and trimmed to the proper shape, should have been taken to a testing machine, and broken, to ascertain its tensilelstrength and ductility, which should be fully equal to the requirements of the specifications. If everything is all right, the iron is now brought to the laying-out table (a low strong bench of heavy plank), a sheet 's placed upon it, and the layer-out commences his work.

First, he tries the sheet with his steel square to see if it is approximately square at the corners; then, with a long straight edge and slate pencil he draws a line through the middle of the

GUIDE POSTS ON THE ENGINEER'S JOURNEY. 117

length of the sheet from end to end; then, at each end and on both sides of this center line, and perpendicular to it, he sets off, with the aid of his square, a distance equal to half the width of the two girth seams as given in the drawings, and through the points thus obtained he draws lines with his straight edge and pencil, which, of course, are parallel to each other and to the center line, and are the center lines for the girth seam rivets. Outside of the lines, and at a distance equal to the width of the lap, he draws lines parallel to them, defining the edges of the sheets with a scriber.

At the ends of the sheets double riveting is required, so he first draws a line with the scriber across the center line and perpendicular to it, just enough within the edge of the sheet as it lies to permit of its being trimmed up properly in the planer, and this is the line of the edge of the end lap. Now from this he lays off a line parallel to it, and at the distance of the width of the lap, which is the center line of the outer line of rivets at that end, and then parallel to this line and at a distance equal to that between the two rows of rivets, he draws the center line for the second row of rivets.

He will now start from the center line of the outer row of rivets and measure along the center line of the sheet a distance, supposing this to be the large course equal to the length of the sheet as found if it just met, 189.67 inches, and through that point he will draw a line at right angles to the center line, which will be the center line for the inner row of rivets at that end, then by setting off a distance equal to that between the two rows of rivets and drawing a line parallel to the one just established he obtains the center line of the outer row of rivets, and then at a distance from this of the width of the lap he draws a parallel line with the scriber for the edge line.

Now he opens his compasses to the desired pitch, and begins at a corner rivet to space off the centers for the holes on the girth, and on reaching the end of the seam will generally find that either he must reset his compass or average the distance over two or three holes. The other girth seam is to be spaced off, and then he spaces the end seams with a different pair of compasses, and marks the centers for all the rivet holes plainly with a center punch. Now if any braces come on the sheet, the places for the rivets which secure them must be marked plainly with a center punch, and any hole to be cut in the sheet must be laid out and its circumference marked plainly by the center punch also. All the shell sheets are laid off in this manner and then sent to be punched.

The next operation is to lay out the head sheets, and one of them is laid on the table, where the center of the sheet, if not already marked, must be found and marked plainly with a center punch, then a circle of the diameter of the boiler must be struck out on it from the center punch mark, and its circumference plainly marked every two or three inches with a center punch as a guide for the flange-turner. If there is to be a flanged manhole in the head, it must be marked out plainly by the center punch, and after allowing width enough inside these marks for the flange, a guide line is drawn, and the head carried to the punch to remove the blank, after which it is carried to the flange fire.

When flanged and cooled, the head is again brought to the table, and two diameters at right angles to each other are scribed through the center. The proper positions for the tube centers are then marked out with a center punch, also the centers for the rivet holes of the brace bars and stays, and the manhole, if it is not to be flanged, or the hand-hole, if there is no manhole, is laid out, and its perimeter marked plainly by a center punch.

Now the head goes to the punching machine, and the rivet holes, centers of the tubes and blanks for holes are all punched, and from there it goes to the drill bench, where the tube holes are drilled out with a peculiar flat drill, with two cutting lips, having a center which is steadied by entering the hole punched at the center of the tube hole; this drill also removes the burr.

Meanwhile, after the shell sheets have been punched, they have been planed to size and edges properly beveled, the inside corners of laps heated and thinned, the sheets rolled and formed up, and then bolted, taken to the riveting machine and riveted.

118   GUIDE POSTS ON THE ENGINEER'S JOURNEY.

When the shell is riveted, the heads are fitted into their proper position and the centers for the rivet holes marked on the flanges, the heads are then taken out, punched, replaced and riveted.

During this time the braces and brace bars have been made by the blacksmith, and they are now riveted in place, as well as the brackets, manhole frames, nozzles and reinforce plates. If there is a dry pipe it generally has to be placed in the boiler before the tubes are set.

The boiler now is ready to be caulked, and the tubes should be put in place, cut to the right length, and set by the expander. The holes should be drilled and tapped for gauge, cocks, and all pipe attachments, and after the hydraulic test has been made, and minor defects made good, it is ready for shipment.

But while all this has been done to the boiler proper, there has also been plenty to do in other departments. The castings for the front have had to be fitted together, doors hung, registers secured in place, liners fitted to doors, anchor and tie bolts made, and, perhaps, also a smoke-pipe, with its hoops and guy rods.

The above is a fair description of the *modus operandi* in the general run of shops. To be sure, of late years, some shops turn a flange on a head at one heat by use of a furnace and hydraulic machine, and by using a multiple drill can drill all the rivet holes for a seam at once, but these tests are not in general use as yet.

In the various operations of punching, shearing, planing and chipping, the quality of the material must become thoroughly known, so that there appears to be no possible excuse for a boiler-maker to plead ignorance if he sends a boiler made of poor stock out of his shop.

For convenience of young engineers, a few of the methods of laying out some of the more usual forms called for in boiler-making are given, as well as some useful definitions.

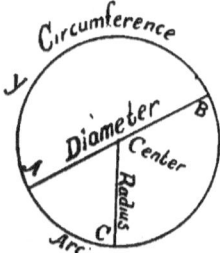

CIRCLE. This is a figure bounded by a curve, called the circumference, every point in which is equally distant from a point within called its center. A diameter is any straight line passing through the center of a circle and bounded by the circumference, as $A B$. It is also just twice the length of the line extending from the center to C, which is called a radius of the circle. An arc is any portion of a circumference, as $A C$, or $A Y B$. The area included between the diameter $A B$ and the portion of the circumference $A Y B$ is called a semi-circle, and $A Y B$ is called a semi-circumference. The length of the circumference of a circle is equal to the diameter multiplied by 3.1416. The area of a circle is equal to the square of the diameter (*i. e.*, the diameter multiplied by itself) multiplied by .7854.

EXAMPLE.—What is the area of a circle whose diameter is six inches?

6 × 6 = 36 (6 squared) ; 36 × 3.1416 = 113.0976 square inches.

```
        36
     188496
      94248
    113.0976
```

To describe a circle of a given diameter (say ten inches in diameter), put one point of the compasses on the rule at any point, and open the legs till the other point strikes a division of the rule just five inches from the first point; now, if you place one point of your compasses on any surface, and revolve the other point of the compasses around the first, letting it only touch the surface lightly, and being careful not to alter the distance between the two points, you will trace the circumference of a circle of ten inches in diameter. The radius with which you describe a circle is always half the diameter of the circle.

TO DIVIDE A LINE INTO TWO EQUAL PARTS.

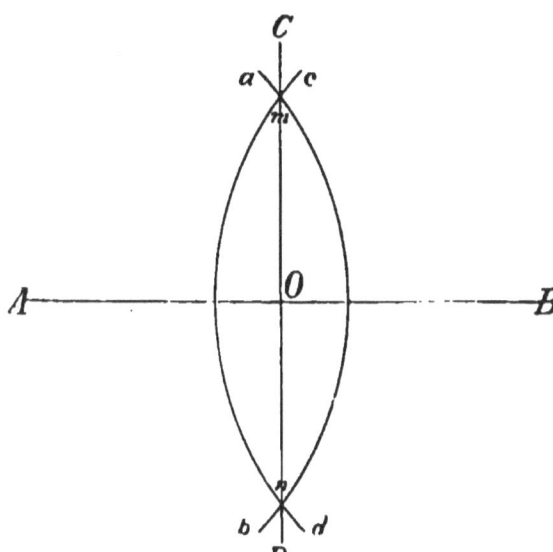

Put one point of the compasses at one extremity, A, and with a radius equal to more than half the line describe the arc $a\ b$; then place the point of the compasses at B, and with the same radius describe the arc $c\ d$, and draw the line $C D$ through the intersections of the arcs at m and n; the line $C D$ divides the line into halves at the point O, and is also perpendicular to it at that point. So again, if it be desired to divide the line into parts, we must the lines A $O B$.

TO DIVIDE A LINE INTO THREE EQUAL PARTS.

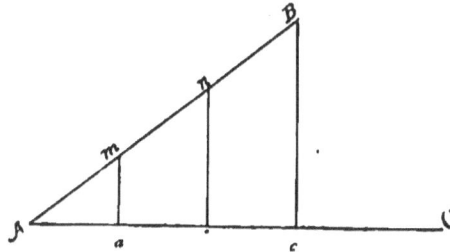

Take the line $A B$ for instance. Draw a line, $A C$, from A, making an angle with $A B$; now lay off three points on $A C$ at equal distances, a, b and c; draw the line $B c$, and draw through the points a and b the lines $a\ m$ and $a\ n$, both parallel to $B c$; the line $A B$ is then divided into three equal parts, $A m$, $m n$, and $n B$.

TO ERECT A PERPENDICULAR TO A LINE AT ONE END.

Let it be required to erect a perpendicular to the line AB at the point B. Assume any point as O, outside of the line AB, and with a radius. OB, describe an indefinite arc, cutting the line AB in C; then from the point C draw an indefinite line through O, and from the point D, in which it cuts the arc, draw a line to B, and this line DB is the perpendicular required.

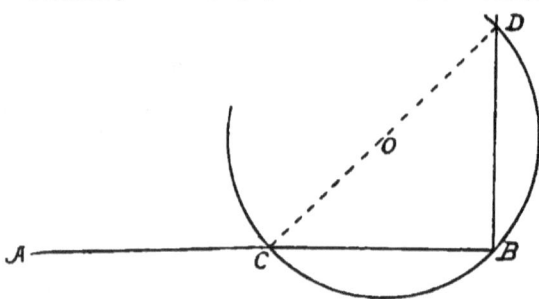

TO ERECT A PERPENDICULAR FROM ANY POINT IN A STRAIGHT LINE.

Let the point C, on the straight line AB, be the one at which it is required to erect a perpendicular. Now, to do this, set one leg of a pair of compasses at C, and then, opening them to any convenient distance, cut the line AB on each side of the point C, as at a and b; then from a as a center, and with a radius greater than aC, describe arcs above and below AB; then, with the same radius, with one point of the compasses on b, describe arcs above and below AB, intersecting those already drawn; now join the points of intersection of these arcs by a straight line, it will pass through C and be perpendicular to AB.

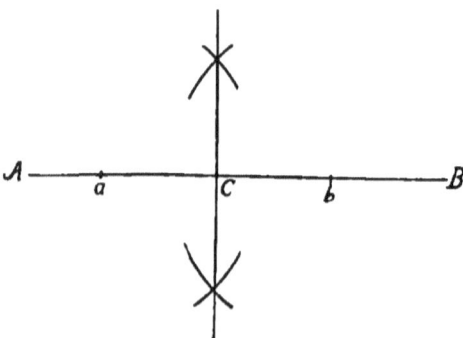

TO DRAW A RECTANGULAR PARALLELOGRAM OF A GIVEN LENGTH AND BREADTH.

Suppose this parallelogram is to have a length of twenty-four inches and a breadth of twelve

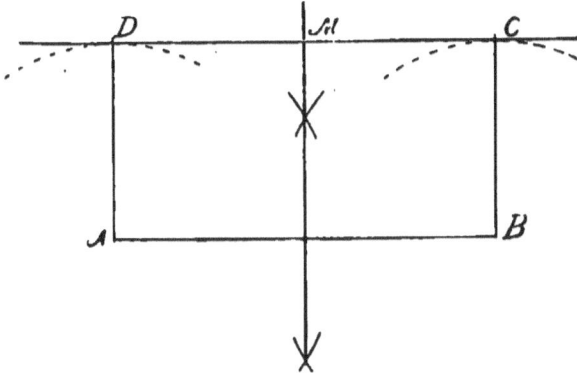

inches. Draw a line, $A\ B$, twenty-four inches long, then take a pair of compasses and open them to twelve inches, place one point on A and describe an indefinite arc, and from B describe an arc also with the same radius; now take a straight edge and draw an indefinite line which will be
parallel to $A\ B$; now divide the line $A\ B$ into two equal parts by erecting a perpendicular at its middle point, and prolong it until it cuts the parallel line in M; now from M measure twelve inches each way, to C and D, and draw lines from D to A, and from C to B. The figure $A\ B\ C\ D$ will be the parallelogram required.

TO DRAW AN ELLIPSE.

This can only be done by means of a string and pins. First mark out the length and breadth

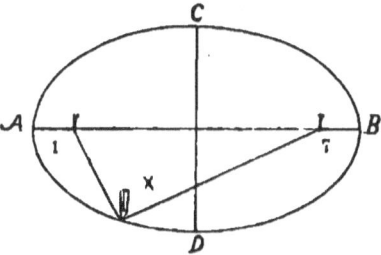

of the figure by two lines at right angles to each other, as $A\ B,\ C\ D$; then divide the major axis $A\ B$ into eight equal parts, and fix two pins at the divisions 1 and 7; now fasten one end of a string to 7, and, sticking a pin temporarily at D, pass the string around it and make the string, when drawn taut, fast to the pin at 1, then remove the temporary pin, put the point of a pencil, as X, inside the loop and describe one-half the ellipse, then shift the position of the string to the other side of the line $A\ B$ and describe the other half.

Pins cannot be used on boiler iron, consequently the ellipse must be drawn on wood or paper and a templet made to scribe the iron with.

The ellipsoidal form is preferred or the manholes of boilers, and on the ollowing page is given two or three methods of laying out these ovals.

TO LAY OUT A MANHOLE 14 BY 10 INCHES.

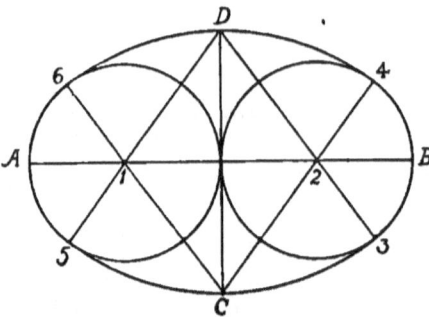

Draw two lines at right angles to each other at their middle points, divide the horizontal line, $A\,B$, which is the longer, into four equal parts, then at the points 1 and 2, as centers, describe circles with the equal radii A 1, and B 2, draw the lines C 4 and C 6 passing through the points 2 and 1, and from C as a center describe, with a radius C 6, the arc 6 D 4; from D draw the lines D 5 and D 3 through the points 1 and 2, and from D as a center, with a radius D 3, describe the arc 3 C 5, and finish the outline of the oval.

TO LAY OUT A MANHOLE 11 BY 15 INCHES.

Draw the horizontal line, $A\,B$, fifteen inches long, and at right angles to its middle point, O, draw the line $C\,D$, eleven inches long, one-half on each side of $A\,B$. Now divide the line $A\,B$ into three equal parts, marking the points of division 1 and 2. Now from $C\,D$ draw lines C 1 and C 2, prolonging them, and draw D 1 and D 2, prolonging them also; now from 1 as a center, with a radius A, describe an arc terminating in the prolonged lines C 1 and D 1 at the points 3 and 5; and from 2, with a radius B 2, describe an arc cutting C 2 and D 2 prolonged in the points 4 and 6; then from the point D, with a radius $C\,D$, describe the arc 5 C 6, and from C, with the same radius, describe the arc 3 D 4, and the oval outline is complete.

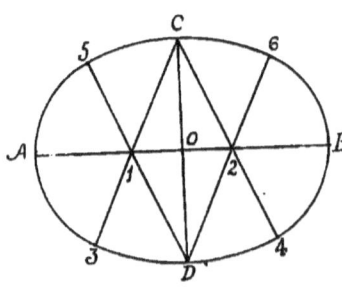

TO LAY OUT A MANHOLE 11 BY 16 INCHES.

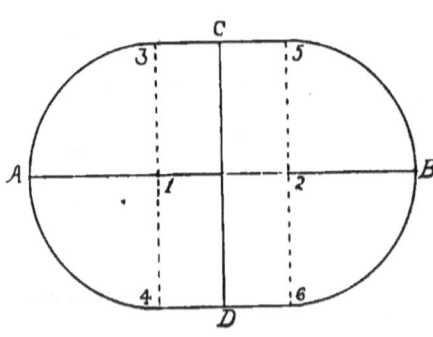

Draw two lines crossing each other at their middle points at right angles. Let the horizontal line be 16 inches in length and measure off from A and B, on the line $A\,B$, 5½ inches towards the center, to the points 1 and 2, then from those points as centers, with a radius of 5½ inches, describe the semicircles 3 A 4, and 5 B 6, and join 3 and 5 by a straight line, and 4 and 6 also by a straight line, and the outline is complete. A manhole made in this shape is not so handsome as either of the others already shown, but it is easier to get in and out of.

TO LAY OUT THE SHEET FOR A DOME.

Let us take the example of an ordinary dome on a cylindrical boiler. We will suppose the dome to be 30 inches high by 30 inches in diameter, and the boiler to have diameter of 60 inches.

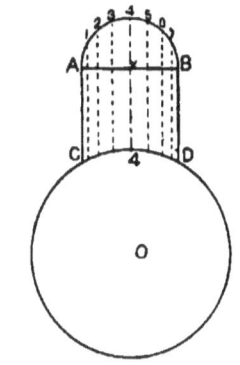

The thickness of the boiler sheets to be ⅜ of an inch, and the dome to be placed on the large course, and consequently the outer circumference of the shell to which the dome must be fitted is curved to a diameter of 60¾ inches. Now draw a circle, which represents a section of the boiler, and draw the figure $A B C D$, which represents the dome, both to full size. Find the center of the line $A B$, which is the diameter of the dome, and from this center, X, describe the semicircle $A\ 1\ 2\ 3\ 4\ 5\ 6\ 7\ B$; divide this semicircle into any number of equal parts (we have taken eight), and from their points of division let fall perpendiculars on the curve of the boiler shell, $C\ 4\ D$. Now draw a horizontal line, $4\ A\ 4\ B$, equal in length to the circumference of the dome, divide it into twice as many parts as the semicircle, and number them as in the diagram; then drop perpendicular lines

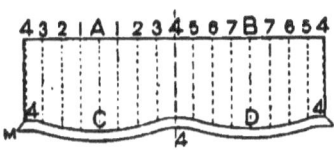

from all these points. Now measure from the line $A B$ in the upper figure the lengths of the perpendiculars let fall to the curve $C\ 4\ D$, and transfer the length to the corresponding perpendiculars let fall from $4\ A\ 4\ B$, marking the points with a center punch; bend a batten and draw in the curve with a slate pencil, and you have the
$4\ B\ D\ 4$, which is the root of the flange, and is the guide for the flange turner. The outer lines, 4 4, at the ends of the sheet, are the center lines of the vertical row of rivets, and at a distance of the width of the required lap from each of these must be drawn a line, as shown, for the edge of the sheet. A distance equal to the required width of the bottom flange must be set off on every perpendicular from the line $4\ B\ D\ 4$, and a curve struck through the points found by aid of a flexible batten, giving the line $M\ 4\ N$. It will be noticed that a little surplus material is left at the corners, M and N, for convenience in working. The sheet now goes to the punch, after being sheared, and is then rolled, formed up, and the vertical seam riveted before it goes to the flange fire, where, by the aid of formers and shapes, the flange-turner forms the flange.

TO LAY OUT THE ENVELOPE OF A FRUSTRUM OF A CONE.

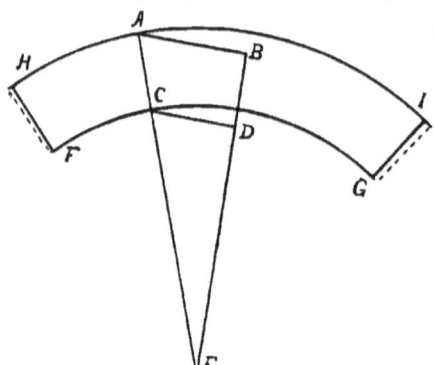

Describe a right angle, $A\,B\,E$; make $B\,D$ equal to the altitude of the frustrum, and draw the line $C\,D$ at right angles to the line $B\,E$; make $A\,B$ equal to one-half the large diameter of the frustrum and $C\,D$ equal to one-half of the small diameter. Draw a line through the points A and C till it cuts the line $B\,D$ extended in E; then with E as a center describe the arcs $H\,I$, passing through A, and $F\,G$, passing through C. Set off the distance equal to the circumference of the base of the frustrum, H to I on the curve, and draw lines from those points towards the center E, cutting the inner curve at F and G, and you have the outline of the envelope. The laps are shown by the exterior dotted lines at $H\,F$ and $I\,G$.

There are many problems in the laying out of special forms that a boiler maker is required to be familiar with, but which would require too much space in any thing but a thorough treatise on the subject; enough, however, has been given to show that a knowledge of practical geometry is essential to a master of the business, as well as practice in the manipulation of tools and materials.

BOILER PLATE TESTS.

All plates intended to be used in the construction of a boiler, must be capable of standing the following tests:

Tensile strain per sq. in., { Lengthwise of the fibre, 45,000 pounds. / Crosswise of the fibre, 40,000 pounds.

All plates must admit of being bent hot without fracture to the following angles: { Lengthwise of the fibre, to 125 degrees. / Across the fibre, to 100 degrees.

All plates must admit of being bent cold, without fracture, to the following angles.
With the fibre, ½″ and 9-16″ plates, to an angle of 35 degrees.
Across the fibre, ½″ and 9-16″ plates, to an angle of 15 degrees.

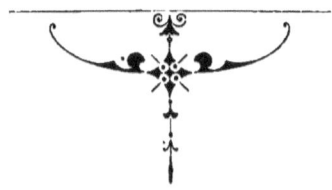

MISCELLANEOUS ITEMS.

The general mode of speaking of a boiler is to call it so many horse-power. In a Return Tubular boiler, fifteen square feet of heating surface is called a horse-power, and twelve square feet in an ordinary Flue or Cylinder boiler.

The heating surface in Return Tubular boiler is taken as the superficial area of all the tubes, and the bottom half of the shell; but in a Flue boiler the superficial area of five-eighths of the shell and all the flues is reckoned as heating surface.

EXAMPLE.—A horizontal Tubular boiler is 4 feet in diameter by 13 feet long, with 45 tubes 3 inches in diameter by 13 feet long. What is its horse-power?

If of 4 feet diameter, it will have 12.566 feet circumference, and half of that is 6.28 feet, which, multiplied by 13 feet, gives 81.64 feet for the half shell area.

Each tube has a diameter of 3 inches, giving 9.42 inches for its circumference, and this, multiplied by 156 (its length in inches), gives 1469.5 square inches, or (dividing by 144) 10.2 square feet for the area of one tube, then multiplying by 45 we have 469 square feet as the area of all the tubes, and adding 81.64 to this, we get 550.64 square feet for the total area. Now dividing this by 15, we have 36.7 horse-power, and we should reckon it as a 35 horse-power boiler.

TO FIND THE WEIGHT PER FOOT IN LENGTH OF ROUND IRON.

Take the diameter in quarter inches, square it, and divide by six.

EXAMPLE.—What is the weight per foot of 2-inch round iron?

2 in. = 8 quarter inches.
8 squared = 8 × 8 = 64.
64 divided by 6 = 10⅔ lbs., the required weight.

WEIGHT OF BOILER IRON.

A cubic foot of wrought iron weighs 480 pounds, consequently a piece one foot square and one inch thick weighs one-twelfth of 480 pounds, or 40, and a plate one foot square and one-sixteenth of an inch thick weighs 2½ lbs. Consequently we have the following

RULE.

Multiply the thickness in sixteenths of an inch by 2½; the result is the weight in pounds per square foot.

EXAMPLE.—The iron of a boiler is $\frac{5}{16}$ of an inch thick. What is the weight per square foot?

5 × 2½ = 12½ lbs., the weight required.

TO REPAIR A FEED OR OTHER WATER PIPE.

Mix a moderately stiff putty from red and white lead with boiled linseed oil, and work into it some hemp chopped into short lengths; lay it over the crack in a moderately thick mass; then wrap some strips of canvass (parceling) round the pipe tightly, overlapping both ends of the crack, and finish by sewing marline hard over the parceling. If the pipe is found to be worn thin and is full of holes, bend a piece of thin sheet copper or tin to the round of the pipe, lay it on the putty, then parcel and serve as before directed.

TO KEEP MACHINERY FROM RUSTING.

Take one ounce of camphor, dissolve it in one pound of melted lard; take off the scum, and mix in as much fine blacklead as will give it iron color. Clean the machinery, and smear it with this mixture. After twenty-four hours, rub clean with a soft linen cloth. It will keep clean for months under ordinary circumstances.

TO BACK OUT BOLTS.

When driving out bolts, without protection for the thread, strike the hardest blow you can with a heavy hammer. Light blows with a small hammer will upset or rivet the bolt.

TO CUT GASKETS.

In cutting rubber for gaskets, etc., have a dish of water handy, and keep wetting the knife blade: it makes the work much easier.

TO CUT A GLASS-GAUGE TUBE.

If a glass-gauge tube is too long, take a three-cornered file and *wet* it; hold the tube in the left hand, with the thumb and forefinger at the place where you wish to cut it; saw it quickly and lightly two or three times with the edge of the file, and it will mark the glass. Now take the tube in both hands, both thumbs being on the opposite side to the mark, and about an inch apart, then try to bend the glass, using your thumbs as fulcrums, and it will break at the mark, which has weakened the tube.

TO FIND OUT WHETHER A PLATE IS BURNED OR CRYSTALLIZED.

Thin, sharp chisel and cut a thin chip for an inch or so; if the chip curls up the iron is

TO PREVENT LAMP CHIMNEYS FROM BREAKING EASILY.

Them in a pot of *cold* water over the fire, and add some common table salt. Boil the water well, and let it cool slowly; then take out the articles and wash them well.

TO CLEAN SMOKY LAMP CHIMNEYS.

Put a teaspoonful of sulphuric acid into about five or six times as much water; then dip into the liquid a piece of flannel tied on a stick, and draw the flannel through the chimney; then rinse in water, and wipe dry.

TO CLEAN VARNISHED PAINT-WORK.

Produce a lot of old tea-grounds, add water, and boil up; apply while hot with a soft piece f flannel, *always rubbing one way*, and rub dry with a soft cloth, or a clean white waste.

TO DRILL HOLES IN GLASS.

A common steel drill, well made and well tempered, is the best tool. The steel should be forged at a low temperature, so as to be sure not to burn it, and then tempered as hard as possible in a bath of salt water that has been well boiled. Such a drill will go through glass very rapidly if kept well moistened with turpentine in which some camphor has been dissolved.

TO SOLDER A BROKEN FILE.

Wet the break with muriate of zinc immediately; then heat a soldering-iron and tin the ends of the file. Heat the file pretty warm, not enough to start the temper, but rather too warm to hold in the hand. When well tinned and hot, press the two pieces together, squeeze out all the solder, and let the file cool. Trim off the joint, and, if well done, the file will break in another place next time. Don't attempt to solder a broken file unless the break is a fresh one.

TO BRAZE SHEET IRON.

Make a solution of borax and water for a flux; mix it with brass spelter and lay it thickly on the iron, and melt over a clear forge fire; remove the work from the fire as soon as the spelter has run into the joint.

TO WIPE A JOINT ON A LEAD PIPE.

Scrape the ends for about one and a quarter inches, and paint with lampblack and oil the part not to be soldered; rub tallow on the parts to be soldered after scraping for a flux. Open one end like a funnel with a wooden tompion, and cut the other end to be joined taper. Hold the ends together in position by clamps, if you have no helper. Take several folds of canvas, or bed ticking, well greased, in your left hand, which is held under the joint, and with a small iron ladle pour molten solder over the joint. With the pad in the left hand catch the solder and press it on the joint; a red-hot soldering iron remelts it, and forms a sound joint, which is finished off with the pad.

ALLOYS.

When mixing different metals, melt the one having the highest melting point first, and then add the others in the order of their melting points, heating them first to prevent their chilling the metal already melted, and stir them with a *wooden rod*. Should the metals tend to volatize, or to form an oxide, keep the surface covered with a layer of fine charcoal. Be sure to skin the surface of the melted metal carefully before pouring.

PLUMBER'S SOLDER.

Melt together two parts of lead and one of tin, by weight.

TINMAN'S SOLDER.

Melt together two parts of tin and one of lead, by weight. Either rosin or muriate of zinc must be used as a flux. When muriate of zinc is used, the joint must be wiped with a wet rag as soon as made to prevent discoloration of the tin.

CEMENT LINING FOR CISTERNS.

Mix 2 parts of powdered brick,
2 parts of quicklime, and
2 parts of wood ashes,
and make into a paste with boiled linseed oil.

CEMENT TO FASTEN IRON TO STONE.

Take 10 parts of fine iron filings,
30 parts of plaster of Paris, and
½ part of sal ammoniac;
mix with weak vinegar to a fluid paste and apply at once.

TO CLEAN BRASS (U. S. Government Method).

Make a mixture of one part common nitric acid and one-half part sulphuric acid in a stove, having also a pail of fresh water and a box of sawdust. Dip the articles into the acid, then soak them in the water, and finally rub them in the sawdust, and they will take on a brilliant color.

If the brass is greasy, it must be first dipped into a strong solution of potash and soda in water, and then rinsed, so that the grease may be removed, leaving the acid free to act.

TO ASCERTAIN THE SAFE WORKING PRESSURE ON A BOILER.

Multiply twice the thickness of the shell by the tensile strength, and divide the product by the diameter of the shell in inches.

STEAM HEATING.

Allow 1 square foot of heating surface in a boiler for every 200 cubic eet of space in a church; n a dwelling-house every 50 cubic feet of space requires 1 square foot of boiler heating surface. The radiators should have 1 square foot of superficial area to every 6 square feet of glass in windows, and 1 square foot for every 80 cubic feet of space to be heated.

One-horse power in a boiler is generally sufficient for 40,000 cubic feet of space for a temperature of 70° Fahrenheit.

TO COOL A HOT BEARING.

Slack off the nuts on the cap; remove the oil cup, if there is one, and supply the journal freely with oil, into which has been stirred the dust obtained by rubbing two bath bricks together. After some time you can set up a little on the nuts, and still continue supplying the mixture. When the nuts have thus been screwed gradually to nearly their former position, supply clear oil in considerable quantity to wash out the journal. Then set up the nuts to place, and it will be a long time before that journal will give trouble.

TO AVOID TEARING MANHOLE GASKETS.

Put a little white lead on that surface of the gasket which rests on the manhole plate, and chalk the other face of the gasket heavily, as also the part of the manhole frame with which it comes into contact. Upon opening the boiler afterwards, the gasket will generally be found to adhere firmly to the plate, and to separate from the frame without tearing.

WEIGHTS AND MEASURES.

TROY WEIGHT.

24 grains.................................	1 pennyweight (dwt.)
20 pennyweights......................	1 ounce (oz.) = 480 grains.
12 ounces	1 pound (lb.) = 5760 grains (gr.).

APOTHECARIES' WEIGHT.

20 grains.................................	1 scruple.
3 scruples...............................	1 dram = 60 grains.
8 drams	1 ounce = 480 grains.
12 ounces	1 pound = 5760 grains.

AVOIRDUPOIS WEIGHT.

27.34375 grains........................	1 dram.
16 drams...............................	1 ounce = 437½ grains.
16 ounces..............................	1 pound = 7000 grains.
28 pounds	1 quarter (qr.).
4 quarters	1 hundredweight (cwt.) = 112 pounds.
20 hundredweights	1 ton (T.) = 2240 pounds.

U. S. LIQUID MEASURE.

4 gills...................................	1 pint (pt.) = 28.875 cubic inches.
2 pints..................................	1 quart (qt.) = 57.750 cubic inches.
4 quarts................................	1 gallon (gal.) = 231 cubic inches.
63 gallons.............................	1 hogshead (hhd.).
2 hogsheads..........................	1 pipe (p.).
2 pipes.................................	1 tun.

APOTHECARIES' FLUID MEASURE.

60 minims (M)........................	1 fluid drachm = 1 fl ʒ.
8 fluid drachms......................	1 fluid ounce = 1 fl ʒ = 480 M.
16 fluid ounces......................	1 fluid pound, or pint = O = 7680 M = 128 fl ʒ
8 pints..................................	1 gallon = Cong = 61240 M = 1024 ʒ = 128 fl ʒ

PRESCRIPTION TABLE.

60 minims, or drops (M or gtts.) .... 1 teaspoonful = 1 ƒ ʒ
4 teaspoonfuls........................... 1 tablespoonful = 4 ƒ ʒ = ½ ƒ ʒ
4 tablespoonfuls......................... 1 wineglassful = 2 ƒ ʒ = ½ gill.
2 wineglassfuls........................... 1 teacupful = 4 ƒ ʒ = 1 gill.
4 wineglassfuls........................... 1 tumblerful = 8 ƒ ʒ = ½ pint = ½ O.

U. S. DRY MEASURE.

2 pints.................. 1 quart (qt.) = 67.2006 cubic inches.
4 quarts .... 1 gallon (gal.) = 8 pts. = 268.8025 cubic inches.
2 gallons................ 1 peck (pk.) = 16 pts. = 8 qts. = 537.605 cubic inches.
4 pecks.................. 1 bushel (bush.) = 64 pts. = 32 qts. = 8 gals. = 2150.42 cubic inches.

LONG MEASURE.

12 inches............................ 1 foot (ft.)
3 feet............................... 1 yard (yd.) = 36 inches.
5½ yards............................. 1 rod (rd.) = 16½ feet.
40 rods.............................. 1 furlong (fur.) = 220 yards = 660 feet.
8 furlongs........................... 1 mile (m.) = 320 rods = 1760 yds. = 5280 ft
3 miles.............................. 1 league (l.) = 960 rods = 5280 yds. = 15840 ft.

SQUARE OR LAND MEASURE.

144 square inches (sq ins.)............. 1 square foot (sq. ft.)
9 square feet......................... 1 square yard (sq. yd.) = 1296 sq. ins.
30¼ square yards...................... 1 square rod (sq. rd.) = 272¼ sq. ft.
40 square rods........................ 1 rood (R.) = 1210 sq. yds. = 10890 sq. ft.
4 Roods............................... 1 acre (A.) = 160 sq. rds. = 4840 sq. yds. = 43560 sq. ft.
640 Acres............................. 1 square mile (sq. m.)

CUBIC OR SOLID MEASURE.

728 cubic inches (cub. ins.)................ 1 cubic or solid foot (cub. ft.).
27 cubic feet.............................. 1 cubic or solid yard (cub. yd.).

MEASURE FOR FIREWOOD, ETC.

1 cord................................... 128 cubic feet.

It is usually piled 4 feet high by 4 feet in breadth, and 8 feet in length. Eight cord feet make a cord—hence a cord foot is 4 feet by 4 feet by 1 foot, or 16 cubic feet.

WEIGHT OF WATER.

1	cubic inch	.03617	pound.
12	cubic inches	.434	pound.
1	cubic foot (salt)	64.3	pounds.
1	cubic foot (fresh)	62.5	pounds.
1	cubic foot	7.48052	U. S. gallons.
1.8	cubic feet	112.0	pounds.
35.84	cubic feet	2240.	pounds.
1	cylindrical inch	.02842	pounds.
12	cylindrical inches	.341	pound.
1	cylindrical foot	49.10	pounds.
1	cylindrical foot	6.0	U. S. gallons.
2.282	cylindrical feet	112.0	pounds.
45.64	cylindrical feet	2240	pounds.
1	imperial gallon	10	pounds.
11.2	imperial gallons	112.0	pounds.
224	imperial gallons	2240	pounds.
1	U. S. gallon	8.355	
13.44	U. S. gallons	112.0	
268.8	U. S. gallons		

NOTE.—The center of pressure of a body of water is at two-thirds the depth from the surface.

To find the pressure in pounds per square inch of a column of water, multiply the height of the column in feet by .434. Every foot elevation is called (approximately) equal to one-half pound pressure per square inch.

MISCELLANEOUS WEIGHTS AND MEASURES.

A point = 1/72 of an inch.
A line = 6 points = 1/12 of an inch.
A palm = 3 inches.
A hand = 4 inches.
A span = 9 inches.
A link = 7.92 inches.
A chain = 100 links = 66 feet = 4 rods.
A fathom = 6 feet.
A nautical mile = 6086 feet and 1/4 in. nearly.
A barrel of flour = 196 pounds.
A barrel of cement = 300 pounds.
A ton of Anthracite coal (broken) = 42 cub. ft.
A ton of bituminous coal = 47 cubic feet.
A stone = 14 pounds.
A load of lime = 32 bushels.
A load of sand = 36 bushels.
A cable's length = 120 fathoms = 720 feet.

An acre = 10 square chains.
A load of bricks = 500 in number.
A cord of wood = 128 cubic feet.
A cord foot = 4 ft. long × 4 ft. high × 1 ft. wide.
A cord = 8 cord feet.
A load of unhewn timber = 40 cubic feet.
A load of squared timber = 50 cubic feet.
A load of inch boards = 600 square feet.
A load of 2-inch planks = 300 square feet.
A cubic foot of tallow = 59 pounds.
A hundred of nails = 120 in number.
A thousand of nails = 1200 in number.
A bushel of sand = 123 pounds.
A bushel of lime 85 pounds.
A ton of coke = 95 cubic feet.

FRENCH WEIGHTS.

REDUCED TO AVOIRDUPOIS WEIGHTS OF 1 POUND=16 OUNCES=7,000 GRAINS.

	Grains.	Ounces.	Pounds.	Tons (2240 lbs.)
Milligramme...	0.01543316	....	....	....
Centigramme...	0.1543316	....	....	....
Decigramme...	1.543316	....	....	....
Gramme...	15.43316	0.0352758	....	....
Decagramme...	....	0.352758	0.02204737	....
Hectogramme...	....	3.52758	0.2204737	....
Kilogramme...	....	35.2758	2.204737	....
Myriogramme...	....	....	22.04737	0.00984258
Quintal...	....	....	220.4737	0.0984258
Tonneau; Millier, or Tonne...	....	....	2204.737	0.984258

FRENCH MEASURES OF LENGTH.

ACCORDING TO U. S. STANDARD.

	U. S. Ins.	U. S. Feet.	U. S. Yards.	U. S. Miles.
Millimeter*...	0.039368	0.003281	....	....
Centimeter†...	0.393685	0.032807	....	....
Decimeter...	3.93685	0.328071	0.109357	....
Meter‡...	39.3685	3.28071	1.09357	....
Decameter...	393.685	32.8071	10.9357	....
Hectometer...	Road Measures.	328.071	109.357	0.0621347
Kilometer...	Road Measures.	3280.71	1093.57	0.621347
Myriameter...	Road Measures.	32807.1	10935.7	6.21347

* Nearly the $\frac{1}{25}$ part of an inch.   † Full ⅖ of an inch.   ‡ Very nearly 3 feet 3⅜ inches, which is too long by only 1 part in 6062.

GUIDE POSTS ON THE ENGINEER'S JOURNEY. 133

FRENCH SQUARE MEASURE.
ACCORDING TO U. S. STANDARD.

	U. S. Sq. Inches.	U. S. Sq. Inches.	U. S. Sq. Yards.	U. S. Acres.	U. S. Miles.
Sq. Millimeter................	0.001549	0.00001076	0.0000012	....	....
Sq. Centimeter................	0.154988	0.00107631	0.0001196	....	....
Sq. Decimeter.................	15.4988	0.10763058	0 0119589	....	....
Sq. Meter or Centiarc..........	1549.88	10.763058	1.195895	0.000247	....
Sq. Decameter or Arc,.........	154988.	1076.3058	119.5895	0.024709	....
Decare (not used)	....	10763.058	1195.895	0.247086	....
Hectare	....	107630.58	11958.95	2.47086	0.0038607
Sq. Kilometer.................	....	10763058.	1195895.	247.086	0.3860716
Sq. Myriameter,...............	....	....	....	24708.6	38.60716

FRENCH CUBIC, OR SOLID MEASURES,
ACCORDING TO U. S. STANDARD.

Milliliter, or cubic centimeter	cub. in. 0.0610165	Liquid, .0084525 gill. Dry, .001816 dry pint.
Centiliter	0.610165	Liquid, .084525 gill. Dry, .01816 dry pint.
Deciliter.............	6.10165	Liquid, .84525 gill = .21131 pint. Dry, .1816 dry pint.
Liter, or cubic decimeter...........	61.0165	Liquid, 1.05656 quart = 2.1131 pints. Dry, .1135 peck = .908 dry qt. = 1.816 dry pt.
Decaliter, or centistère	610.165 cub. feet. 0.353105	Liquid, 2.64141 U. S. gallons. Dry, .283742 bush. = 1.135 pks. = 9.08 dry qts.
Hectoliter, or decistère	3.53105	Liquid, 26.4141 U. S. gallons. Dry, 2.83742 bushels.
Kiloliter, or cubic meter, or store...	35.3105	Liquid, 264.141 U. S. gal...... cub. yds., 1.3078 Dry, 28.3742 bushels......
Myrioliter or décastère.............	353.105	Liquid, 2641.41 U. S. gal...... cub. yds., 13.078 Dry, 283.742 bushels.......

EQUIVALENTS OF THE DECIMAL PARTS OF A POUND (16 OUNCES) IN OUNCES.

Decimals.	Ounces.	Decimals.	Ounces.	Decimals.	Ounces.
0.0625	1	0.375	6	0.6875	11
0.09375	1½	0.40625	6½	0.71875	11½
0.125	2	0.4375	7	0.75	12
0.15625	2½	0.46875	7½	0.78125	12½
0.1875	3	0.5	8	0.8125	13
0.21875	3½	0.53125	8½	0.84375	13½
0.25	4	0.5625	9	0.875	14
0.28125	4½	0.59375	9½	0.90625	14½
0.3125	5	0.625	10	0.9375	15
0.34375	5½	0.65625	10½	0.96875	15½

WEIGHT OF CAST-IRON BALLS,
FROM 1 INCH TO 12 INCHES DIAMETER.

Diameter.	Weight.	Diameter.	Weight.	Diameter.	Weight.	Diameter.	Weight.	Diameter.	Weight.
In.	Lbs.	In.	Lbs.	In.	Lbs.	In.	Lbs.	In.	Lbs.
1	0.136	3½	5.84	6	29.45	8½	83.73	11	181.48
1½	0.460	4	8.72	6½	37.44	9	99.4	11½	207.37
2	1.09	4½	12.42	7	46.76	9½	116.9	12	235.62
2½	2.13	5	17.04	7½	57.52	10	136.35		
3	3.68	5½	22.68	8	69.81	10½	157.84		

WEIGHT OF A LINEAL FOOT OF FLAT, BAR AND HOOP IRON IN POUNDS.

Thickness in Inches.	BREADTH IN INCHES.									
	3½	3	2¾	2½	2¼	2	1¾	1½	1¼	1	¾
⅛	1.47	1.26	1.15	1.05	.094	.084	.073	.063	.052	.042	.031
3/16	2.20	1.89	1.73	1.57	1.41	1.26	1.10	.094	.078	.063	.047
¼	2.94	2.52	2.31	2.10	1.89	1.68	1.47	1.26	1.05	.084	.063
⅜	4.41	3.78	3.46	3.15	2.83	2.52	2.20	1.89	1.57	1.26	.094
½	5.88	5.04	4.62	4.20	3.78	3.36	2.94	2.22	2.10	1.68	1.26
⅝	7.35	6.30	5.77	5.25	4.72	4.20	3.67	3.15	2.62	2.10	1.57
¾	8.82	7.56	6.93	6.30	5.66	5.04	4.41	3.78	3.15	2.52	
⅞	10.29	8.82	8.08	7.35	6.61	5.88	5.14	4.41	3.67	2.94	
1 in.	11.76	10.08	9.24	8.40	7.56	6.72	5.87	5.04	4.20		

WEIGHT OF A SQUARE FOOT OF PLATE IRON IN POUNDS.

Thickness in parts of an inch	⅛	3/16	¼	5/16	⅜	7/16	½	9/16	⅝	11/16	¾
Weight in pounds..	5	7½	10	12½	15	17½	20	22½	25	27½	30

WEIGHT OF A SQUARE FOOT OF SHEET IRON IN POUNDS.

Number on wire gauge	1	2	3	4	5	6	7	8	9	10	11
And weight in pounds	12.5	12	11	10	9	8	7.5	7	6	5.68	5
Number on wire gauge	12	13	14	15	16	17	18	19	20	21	22
And weight in pounds	4.62	4.32	4	3.95	3	2.5	2.18	1.93	1.62	1.5	1.37

WEIGHT OF A SQUARE FOOT OF SHEET AND PLATE COPPER IN POUNDS

Number on wire gauge	12	13	14	15	16	17	18	19	20	21	22
And weight in pounds.	5.08	4.34	3.60	3.27	2.90	2.52	2.15	1.97	1.78	1.62	1.45
Thickness in parts of an inch......	$\frac{3}{32}$	$\frac{3}{16}$	$\frac{7}{32}$	$\frac{1}{4}$	$\frac{9}{32}$	$\frac{5}{16}$	$\frac{11}{32}$	$\frac{3}{8}$	$\frac{13}{32}$	$\frac{7}{16}$	$\frac{1}{2}$
Weight in pounds....	7.26	8.71	10.16	11.61	13.07	14.52	15.97	17.41	18.87	20.32	23.22

NUMBER OF U. S. GALLONS (231 CUBIC INCHES) IN 1 FOOT LENGTH OF PIPE OF DIFFERENT DIAMETERS.

Diam. in inches.	Gallons.	Diam. in inches.	Gallons.	Diam. in inches.	Gallons.
$\frac{3}{4}$	.0230	$3\frac{1}{2}$	.5000	10	4.081
1	.0408	4	.6528	11	4.937
$1\frac{1}{4}$	.0638	$4\frac{1}{2}$	.8263	12	5.876
$1\frac{1}{2}$	.0918	5	1.020	13	6.895
$1\frac{3}{4}$	.1250	6	1.469	14	.997
2	.1632	7	1.999	15	9.180
$2\frac{1}{2}$	.2550	8	2.611	16	10.44
3	.3673	9	3.305	18	13.22

NUMBER OF U. S. GALLONS (231 CUBIC INCHES) CONTAINED IN CIRCULAR TANKS.

Depth in feet.	Diam. Inches.	1 Gals.	2 Gals.	3 Gals.	4 Gals.	5 Gals.	6 Gals.	7 Gals.	8 Gals.	9 Gals.	10 Gals.	Diam. Inches.	Depth in feet.
	20	16.32	32.64	48.96	65.28	81.60	97.92	114.24	130.56	146.88	163.20	20	
	24	23.50	47.00	70.50	94.00	117.50	141.00	164.50	188.00	211.50	235.00	24	
	26	27.58	55.16	82.74	110.32	137.90	165.48	193.06	220.64	248.22	275.80	26	
	28	31.99	63.98	95.97	127.96	159.95	191.94	223.93	255.92	288.91	319.90	28	
	30	36.72	73.44	110.16	146.88	183.60	220.32	257.04	293.76	330.48	367.20	30	
	36	52.88	105.76	158.64	211.52	264.40	317.28	370.16	423.04	475.92	528.80	36	
	42	71.96	143.92	215.88	287.84	359.80	431.76	503.72	575.68	647.64	719.60	42	
	45	82.62	165.24	247.86	330.48	413.10	495.72	578.34	660.96	743.58	826.20	45	
	48	94.02	188.04	282.06	376.08	470.10	564.12	658.14	752.16	846.18	940.20	48	
	50	102.00	204.00	306.00	408.00	510.00	612.0	714.0	816.0	918.0	1020.0	50	
	54	119.0	238.00	357.00	476.00	595.00	714.0	833.0	952.0	1071.0	1190.0	54	
	60	146.9	293.8	440.70	587.6	734.5	881.4	1028.3	1175.2	1322.1	1469.0	60	
	66	177.7	355.4	533.10	710.8	888.5	1066.2	1243.9	1421.6	1599.2	1777.0	66	
	72	211.5	433.0	534.50	846.0	1057.5	1269.0	1480.5	1692.0	1903.5	2115.0	72	
	84	287.8	575.6	863.4	1151.2	1439.0	1726.8	2014.6	2302.4	2590.2	2878.0	84	

CONTENTS OF CIRCULAR TANKS IN U. S. GALLONS (231 CUBIC INCHES).

Depth. Diam.	8 Gals.	9 Gals.	10 Gals.	11 Gals.	12 Gals.	13 Gals.	14 Gals.	15 Gals.	16 Gals.	17 Gals.	18 Gals.	19 Gals.	20 Gals.
10	4,354	4,898	5,442	5,986	6,531	7,075	7,619	8,163	8,707	9,252	9,796	10,340	10,884
12	6,767	7,614	8,460	9,306	10,152	10,998	11,844	12,690	13,536	14,362	15,228	16,074	16,920
14	9,212	10,363	11,515	12,666	13,818	14,969	16,121	17,273	18,424	19,576	20,726	21,878	23,030
16	12,031	13,535	15,040	16,544	18,048	19,552	21,056	22,560	24,064	25,568	27,070	28,575	30,080
18	15,228	17,132	19,034	20,939	22,842	24,746	26,649	28,553	30,456	32,360	34,264	36,166	38,068
20	18,800	21,150	23,499	25,850	28,200	30,550	32,901	35,251	37,601	39,951	42,300	44,649	46,998
22	22,748	25,592	28,435	31,277	34,123	36,966	39,810	42,653	45,497	48,340	51,184	54,027	56,870
24	27,068	30,456	33,841	37,225	40,607	43,993	47,377	50,761	54,145	57,529	60,912	64,297	67,682
26	31,777	35,744	39,716	43,687	47,659	51,628	55,602	59,574	63,545	67,517	71,488	75,460	79,432
28	36,849	41,453	46,061	50,667	55,273	59,879	64,481	69,092	73,698	78,304	82,904	87,513	92,122
30	42,301	47,588	52,876	58,164	63,451	68,739	74,027	79,310	84,602	89,890	95,176	100,464	105,752
32	48,129	54,140	60,161	66,178	72,194	78,210	84,226	90,242	96,253	102,275	108,280	114,301	120,322
34	54,335	61,125	67,917	74,708	81,500	88,292	95,083	101,875	108,667	115,451	122,250	129,042	135,834
36	60,915	68,530	76,144	83,758	91,373	98,987	106,602	114,216	121,831	129,445	137,059	144,674	152,288
38	67,870	76,354	84,838	93,321	101,805	110,289	118,773	127,257	135,740	144,224	152,708	161,192	169,675
40	75,202	84,603	94,003	103,403	112,804	122,204	131,604	141,005	150,405	159,805	169,205	178,606	188,006

Depth and diameters are both in feet.   1 cubic foot = 7.8405 gallons.

RELATIVE VALUE OF NON-CONDUCTORS.
(C. E. EMERY.)

Non-Conductors.	Value.
Wool Felt	1.000
Mineral Wool, No. 2	.632
" " with Tar	.715
Sawdust	.680
Mineral Wool, No. 1	.676
Charcoal	.632
Pinewood, across fiber	.553
Loam, dry and open	.550
Slacked Lime	.480
Gas-house Carbon	.470
Asbestos	.363
Coal Ashes	.345
Coke, in lumps	.277
Air Space, undivided	.186

IGNITION POINTS OF VARIOUS SUBSTANCES.

Phosphorus ignites at	150° Fahr.
Sulphur " "	500° "
Wood " "	800° "
Coal " "	1000° "
Lignite, in the form of dust, ignites at	150° "
Cannel Coal, " " "	200° "
Coking Coal, " " "	250° "
Anthracite, " " "	300° "

TABLE OF THE STRONGEST FORM AND PROPORTION OF RIVETED JOINTS.

Thickness of Plate.	Diameter of Rivet.	Length of Rivet.	Pitch.	Lap.
1/16 inch.	3/8 inch.	0.85 inches.	1.14 inches.	1.14 inches.
1/4 "	1/2 "	1.12 "	1.5 "	1.5 "
1/16 "	5/8 "	1.39 "	1.55 "	1.76 "
3/8 "	3/4 "	1.68 "	1.87 "	2.1 "
1/2 "	3/4 "	2.25 "	2.00 "	2.25 "
5/8 "	1 "	2.82 "	2.5 "	2.82 "
3/4 "	1 1/4 "	3.37 "	3.0 "	3.37 "

CIRCUMFERENCES AND AREAS OF CIRCLES.

Diam.	Circum.	Area.	Diam.	Circum.	Area.	Diam.	Circum.	Area.
Ins.	Ins.	Sq. ins.	Ins.	Ins.	Sq. ins.	Ins.	Ins.	Sq. ins.
1/16	.19635	.00307	3 ins.	9.4248	7.0686	6 ins.	18.849	28.274
1/8	.3927	.0122	1/8	9.8175	7.6699	1/8	19.242	29.464
1/4	.7854	.0490	1/4	10.210	8.2957	1/4	19.635	30.679
3/8	1.1781	.1104	3/8	10.602	8.9462	3/8	20.027	31.919
1/2	1.5708	.1963	1/2	10.995	9.6211	1/2	20.420	33.183
5/8	1.9635	.3068	5/8	11.388	10.320	5/8	20.813	34.471
3/4	2.3562	.4417	3/4	11.781	11.044	3/4	21.205	35.784
7/8	2.7489	.6013	7/8	12.173	11.793	7/8	21.598	37.122
1 in.	3.1416	.7854	4 ins.	12.566	12.566	7 ins.	21.991	38.484
1/8	3.5343	.9940	1/8	12.959	13.364	1/8	22.383	39.871
1/4	3.9270	1.2271	1/4	13.351	14.186	1/4	22.776	41.282
3/8	4.3197	1.4848	3/8	13.744	15.033	3/8	23.169	42.718
1/2	4.7124	1.7671	1/2	14.137	15.904	1/2	23.562	44 1/8
5/8	5.1051	2.0739	5/8	14.529	16.800	5/8	23.954	45.663
3/4	5.4978	2.4052	3/4	14.922	17.720	3/4	24.347	47.173
7/8	5.8905	2.7611	7/8	15.315	18.665	7/8	24.740	48.707
2 ins.	6.2832	3.1416	5 ins.	15.708	19.635	8 ins.	25.132	50.265
1/8	6.6759	3.5465	1/8	16.100	20.629	1/8	25.515	51.848
1/4	7.0686	3.9760	1/4	16.493	21.647	1/4	25.918	53.456
3/8	7.4613	4.4302	3/8	16.886	22.690	3/8	26.310	55.088
1/2	7.8540	4.9087	1/2	17.278	23.758	1/2	26.703	56.745
5/8	8.2467	5.4119	5/8	17.671	24.850	5/8	27.096	58.426
3/4	8.6394	5.9395	3/4	18.064	25.967	3/4	27.489	60.132
7/8	9.0321	6.4918	7/8	18.457	27.108	7/8	27.881	61.862

GUIDE POSTS ON THE ENGINEER'S JOURNEY. 141

CIRCUMFERENCES AND AREAS OF CIRCLES.—Continued.

Diam.	Circum.	Area.	Diam.	Circum.	Area.	Diam.	Circum.	Area.
Ins.	Ins.	Sq. ins.	Ins.	Ins.	Sq. ins.	Ins.	Ins.	Sq. ins.
9 ins.	28.274	63.617	12 ins.	37.699	113.097	15 ins.	47.124	176.715
⅛	28.667	65.396	⅛	38.091	115.466	⅛	47.516	179.672
¼	29.059	67.200	¼	38.484	117.859	¼	47.909	182.654
⅜	29.452	69.029	⅜	38.877	120.276	⅜	48.302	185.661
½	29.845	70.882	½	39.270	122.718	½	48.694	188.692
⅝	30.237	72.759	⅝	39.662	125.184	⅝	49.087	191.748
¾	30.630	74.662	¾	40.055	127.676	¾	49.480	194.828
⅞	31.023	76.588	⅞	40.448	130.192	⅞	49.872	197.93.
10 ins.	31.416	78.540	13 ins	40.840	132.732	16 ins.	50.265	201.4 2
⅛	31.808	80.515	⅛	41.233	135.297	⅛	50.658	204.216
¼	32.201	82.516	¼	41.626	137.886	¼	51.051	207.394
⅜	32.594	84.540	⅜	42.018	140.500	⅜	51.443	210.597
½	32.986	86.590	½	42.411	143.137	½	51.836	213.825
⅝	33.379	88.664	⅝	42.804	145.802	⅝	52.229	217.077
¾	33.772	90.762	¾	43.197	148.489	¾	52.621	220.353
⅞	34.164	92.885	⅞	43.589	151.201	⅞	53.014	223.654
11 ins.	34.557	95.033	14 ins.	43.982	153.938	17 ins.	53.407	226.980
⅛	34.950	97.205	⅛	44.375	156.699	⅛	53.799	230.33.
¼	35.343	99.402	¼	44.767	159.485	¼	54.192	233.705
⅜	35.735	101.623	⅜	45.160	162.295	⅜	54.585	237.104
½	36.128	103.869	½	45.553	165.130	½	54.978	240.52
⅝	36.521	106.139	⅝	45.945	167.989	⅝	55.370	243.977
¾	36.913	108.434	¾	46.338	170.873	¾	55.763	247.450
⅞	37.306	110.753	⅞	46.731	173.782	⅞	56.156	250.947

CIRCUMFERENCES AND AREAS OF CIRCLES.—Continued.

Diam.	Circum.	Area.	Diam.	Circum.	Area.	Diam.	Circum.	Area.
Ins.	Ins.	Sq. ins.	Ins.	Ins.	Sq. ins.	Ins.	Ins.	Sq. ins.
18 ins.	56.548	254.469	21 ins.	65.973	346.361	24 ins.	75.398	452.390
⅛	56.941	258.016	⅛	66.366	350.497	⅛	75.791	457.115
¼	57.334	261.587	¼	66.759	354.657	¼	76.183	461.864
⅜	57.726	265.182	⅜	67.151	358.841	⅜	76.576	466.638
½	58.119	268.803	½	67.544	363.051	½	76.969	471.436
⅝	58.512	272.447	⅝	67.937	367.284	⅝	77.361	476.259
¾	58.905	276.117	¾	68.329	371.543	¾	77.754	481.106
⅞	59.297	279.811	⅞	68.722	375.826	⅞	78.147	485.978
19 ins.	59.690	283.529	22 ins.	69.115	380.133	25 ins.	78.540	490.875
⅛	60.083	287.272	⅛	69.507	384.465	⅛	78.932	495.796
¼	60.475	291.039	¼	69.900	388.822	¼	79.325	500.741
⅜	60.868	294.831	⅜	70.293	393.203	⅜	79.718	505.711
½	61.261	298.648	½	70.686	397.608	½	80.110	510.706
⅝	61.663	302.489	⅝	71.078	402.038	⅝	80.503	515.725
¾	62.046	306.355	¾	71.471	406.493	¾	80.896	520.769
⅞	62.439	310.245	⅞	71.864	410.972	⅞	81.288	525.837
20 ins.	62.832	314.160	23 ins.	72.256	415.476	26 ins.	81.681	530.930
⅛	63.224	318.099	⅛	72.649	420.004	⅛	82.074	536.047
¼	63.617	322.063	¼	73.042	424.557	¼	82.467	541.189
⅜	64.010	326.051	⅜	73.434	429.135	⅜	82.859	546.356
½	64.402	330.064	½	73.827	433.731	½	83.252	551.547
⅝	64.795	334.101	⅝	74.220	438.363	⅝	83.645	556.762
¾	65.188	338.163	¾	74.613	443.014	¾	84.037	562.002
⅞	65.580	342.250	⅞	75.005	447.699	⅞	84.430	567.267

CIRCUMFERENCES AND AREAS OF CIRCLES.—Continued.

Diam.	Circum.	Area.	Diam.	Circum.	Area.	Diam.	Circum.	Area.
Ins.	Ins.	Sq. ins.	Ins.	Ins.	Sq. ins.	Ins.	Ins.	Sq. ins.
27 ins.	84.823	572.556	30 ins.	94.248	706.860	33 ins.	103.672	855.30
⅛	85.215	577.870	⅛	94.640	712.762	⅛	104.055	861.79
¼	85.608	583 208	¼	95.033	718.690	¼	104 458	868.30
⅜	86.001	588.571	⅜	95.426	724.641	⅜	104.850	874 84
½	86.394	593.958	½	95.818	730.618	½	105.243	881.41
⅝	86.786	599.370	⅝	96.211	736.619	⅝	105.636	888.00
¾	87.179	604.807	¾	96.604	742.644	¾	106.029	894.61
⅞	87.572	610.268	⅞	96.996	748.694	⅞	106.421	901.25
28 in.	87.964	615.753	31 ins.	97.389	754.769	34 ins.	106.814	907.92
⅛	88.357	621.263	⅛	97 782	760.868	⅛	107.207	914.61
¼	88.750	626.798	¼	98.175	766.992	¼	107.599	921.32
⅜	89.142	632.357	⅜	98.567	773.140	⅜	107.992	928.06
½	89.535	637.941	½	98.968	779 313	½	108.385	934.82
⅝	89.928	643.594	⅝	99.353	785.510	⅝	108.777	941.60
¾	90.321	649.182	¾	99.745	791.732	¾	109.170	948.41
⅞	90.713	654.839	⅞	100.138	797.978	⅞	109.563	955.25
29 ins.	91.106	660.521	32 ins.	100.531	804.249	35 ins.	106.956	962.11
⅛	91.499	666.277	⅛	100.924	810.545	⅛	110.349	968.99
¼	91.891	671 958	¼	101.316	816.865	¼	110.741	975.90
⅜	92.284	677.714	⅜	101.709	823.209	⅜	111.134	982.84
½	92.677	683.494	½	102.102	829.578	½	111.526	989.80
⅝	93.069	689.298	⅝	102.494	835.972	⅝	111.919	996.78
¾	93.462	695.128	¾	102 887	842.390	¾	112.312	1003.78
⅞	93.855	700.981	⅞	103.280	848.833	⅞	112.704	1010.82

CIRCUMFERENCES AND AREA OF CIRCLES.—Continued.

Diam	Circum.	Area.	Diam.	Circum.	Area.	Diam.	Circum.	Area
Ins.	Ins.	Sq. ins.	Ins.	Ins.	Sq. ins	Ins.	Ins.	Sq. ins.
36 ins.	113.097	1017.87	39 ins.	122.522	1194.59	42 ins.	131 947	1385.44
⅛	113.490	1024.95	⅛	122.915	1202.26	⅛	132.339	1393.70
¼	113.883	1032 06	¼	123.307	1209 95	¼	132.732	1401.98
⅜	114.275	1039 19	⅜	123.700	1217.67	⅜	133.125	1410.29
½	114.668	1046.35	½	124.093	1225.42	½	133.518	1418.62
⅝	115.061	1053.52	⅝	124 485	1233 18	⅝	133.910	1426.98
¾	115.453	1060 73	¾	124 878	1240.98	¾	134.303	1435.36
⅞	115.846	1067.95	⅞	125.271	1248.79	⅞	134.696	1443.77
37 ins.	116.239	1075.21	40 ins.	125.664	1256.64	43 ins.	135.088	1452.20
⅛	116 631	1082.48	⅛	126.056	1264.50	⅛	135.481	1460.65
¼	117 024	1089.79	¼	126.449	1272.39	¼	135.874	1469.13
⅜	117.417	1 97.11	⅜	126.842	1280.31	⅜	136.266	1477.63
½	117 810	1104 46	½	127.234	1288 25	½	136.659	1486.17
⅝	118.202	1111.84	⅝	127.627	1296.21	⅝	137.052	1494.72
¾	118.595	1119.24	¾	128.020	1304.20	¾	137 445	1503 30
⅞	118.988	1126.66	⅞	128.412	1312.21	⅞	137 827	1511.90
38 ins.	119 380	1134.11	41 ins.	128.805	1320.25	44 ins.	138 230	1520.53
⅛	119 773	1141.59	⅛	129.198	1328 32	⅛	138 623	1529.18
¼	120.166	1149.08	¼	129.591	1336.40	¼	139 015	1537.86
⅜	120.558	1156.61	⅜	129.983	1344.51	⅜	139.408	1546.55
½	120.951	1164.15	½	130.376	1352.65	½	139.801	1555.28
⅝	121.344	1171.73	⅝	130.769	1360.81	⅝	140.193	1564 03
¾	121.737	1179.32	¾	131.161	1369.60	¾	140.586	1572.81
⅞	122.129	1186.94	⅞	131.554	1377.21	⅞	140.979	1581.61

CIRCUMFERENCES AND AREA OF CIRCLES.—Continued.

Diam.	Circum.	Area.	Diam.	Circum.	Area.	Diam.	Circum.	Area.
Ins.	Ins.	Sq. ins.	Ins.	Ins.	Sq. ins.	Ins.	Ins.	Sq. ins.
45 ins.	141.372	1590.43	48 ins.	150.796	1809.56	52 ins.	163.363	2123.72
⅛	141.764	1599.28	⅛	151.189	1818.99	¼	164.148	2144.19
¼	142.157	1608.15	¼	151.582	1828.46	½	164.934	2164.75
⅜	142.550	1617.04	⅜	151.974	1837.93	¾	165.719	2185.42
½	142.942	1625.97	½	152.367	1847.45	53 ins.	166.504	2206.18
⅝	143.335	1634.92	⅝	152.760	1856.99	¼	167.290	2227.05
¾	143.728	1643.89	¾	153.153	1866.55	½	168.075	2248.01
⅞	144.120	1652.88	⅞	153.545	1876.13	¾	168.861	2269.06
46 ins.	144.513	1661.90	49 ins.	153.938	1885.74	54 ins.	169.646	2290.22
⅛	144.906	1670.95	⅛	154.331	1895.37	¼	170.431	2311.48
¼	145.299	1680.01	¼	154.723	1905.03	½	171.217	2332.83
⅜	145.691	1689.10	⅜	155.116	1914.70	¾	172.002	2354.28
½	146.084	1698.23	½	155.509	1924.42	55 ins.	172.788	2375.83
⅝	146.477	1707.37	⅝	155.901	1934.15	¼	173.573	2397.48
¾	146.869	1716.54	¾	156.294	1943.91	½	174.358	2419.22
⅞	147.262	1725.73	⅞	156.687	1953.69	¾	175.144	2441.07
47 ins.	147.655	1734.94	50 ins.	157.080	1963.50	56 ins.	175.929	2463.01
⅛	148.047	3744.18	¼	157.865	1983.18	¼	176.715	2485.05
¼	148.440	1753.45	½	158.650	2002.96	½	177.500	2507.19
⅜	148.833	1762.73	¾	159.436	2022.84	¾	178.285	2529.42
½	149.226	1772.05	51 ins.	160.221	2042.82	57 ins.	179.071	2551.76
⅝	149.618	1781.39	¼	161.007	2062.90	¼	179.856	2574.19
¾	150.011	1790.76	½	161.792	2083.07	½	180.642	2596.72
⅞	150.404	1800.14	¾	162.577	2103.35	¾	181.427	2619.35

CIRCUMFERENCES AND AREAS OF CIRCLES.—Continued.

Diam.	Circum.	Area.	Diam.	Circum.	Area.	Diam.	Circum.	Area.
Ins.	Ins.	Sq. ins.	Ins.	Ins.	Sq. ins.	Ins.	Ins.	Sq. ins.
58 ins.	182.212	2642.08	64 ins.	201.062	3216.99	70 ins.	219.912	3848.46
¼	182.998	2664.91	¼	201.847	3242.17	¼	220.697	3875.99
½	183.783	2687.83	½	202.683	3267.46	½	221.482	3903.63
¾	184.569	2710.85	¾	203.418	3292.83	¾	222.268	3931.36
59 ins.	185.354	2733.97	65 ins.	204.204	3318.31	71 ins.	223.053	3959.20
¼	186.139	2757.19	¼	204.989	3343.88	¼	223.839	3987.13
½	186.925	2780.51	½	205.774	3369.56	½	224.624	4015.16
¾	187.710	2803.92	¾	206.560	3395.33	¾	225.409	4043.28
60 ins.	188.496	2827.44	66 ins.	207.345	3421.20	72 ins.	226.195	4071.51
¼	189.281	2851.05	¼	208.131	3447.16	¼	226.980	4099.83
½	190.066	2874.76	½	208.916	3473.23	½	227.766	4128.25
¾	190.852	2898.56	¾	209.701	3499.39	¾	228.551	4156.77
61 ins.	191.637	2922.47	67 ins.	210.487	3525.66	73 ins.	229.336	4185.39
¼	192.423	2946.47	¼	211.272	3552.01	¼	230.122	4214.11
½	193.208	2970.57	½	212.058	3578.47	½	230.907	4242.92
¾	193.993	2994.77	¾	212.843	3605.03	¾	231.693	4271.83
62 ins.	194.779	3019.07	68 ins.	213.628	3631.68	74 ins.	232.478	4300.85
¼	195.564	3043.47	¼	214.414	3658.44	¼	233.263	4329.95
½	196.350	3067.96	½	215.199	3685.29	½	234.049	4359.16
¾	197.135	3092.56	¾	215.985	3712.24	¾	234.834	4388.47
63 ins.	197.920	3117.25	69 ins.	216.770	3739.28	75 ins.	235.620	4417.87
¼	198.706	3142.04	¼	217.555	3766.43	¼	236.405	4447.37
½	199.491	3166.92	½	218.341	3793.67	½	237.190	4476.97
¾	200.277	3191.91	¾	219.126	3821.02	¾	237.976	4506.6,

CIRCUMFERENCES AND AREAS OF CIRCLES.—Continued.

Diam	Circum.	Area.	Diam.	Circum.	Area.	Diam.	Circum.	Area.
Ins.	Ins.	Sq. ins.	Ins.	Ins.	Sq. ins.	Ins.	Ins.	Sq. ins.
76 ins.	238.761	4536.37	84 ins.	263.894	5541.78	96	301.594	7238.25
¼	239.547	4566.36	½	265.465	5607.95	97	304.735	7389.83
½	240.332	4596.35	85 ins.	267.036	5674.51	98	307.877	7542.98
¾	241.117	4626.44	½	268.606	5741.47	99	311.018	7784.10
77 ins.	241.903	4656.63	86 ins.	270.177	5808.81	100	314.159	7854.00
¼	242.688	4686.92	½	271.748	5876.55	101	317.301	8011.86
½	243.474	4717.30	87 ins.	273.319	5944.66	102	320.442	8171.30
¾	244.259	4747.79	½	274.890	6013.21	103	323.584	8332.31
78 ins.	245.044	4778.37	88 ins.	276.460	6082.13	104	326.725	8494.88
¼	245.830	4809.05	½	278.031	6151.44	105	329.867	8659.03
½	246.615	4839.83	89 ins.	279.602	6221.15	106	333.009	8824.75
¾	247.401	4870.70	½	281.173	6291.25	107	336.150	8992.04
79 ins.	248.186	4901.68	90 ins	282.744	6361.74	108	339.292	9160.90
¼	248.971	4932.75	½	284.314	6432.62	109	342.433	9331.34
½	249.757	4963.92	91 ins.	285.885	6503.89	110	345.575	9503.34
¾	250.542	4995.19	½	287.456	6573.56			
80 ins.	251.328	5026.56	92 ins.	289.027	6647.62			
½	252.898	5089.58	½	290.598	6720.07			
81 ins.	254.469	5153.00	93 ins.	292.168	6792.92			
½	256.040	5216.82	½	293.739	6866.16			
82 ins.	257.611	5281.02	94 ins.	295.310	6939.79			
½	259.182	5345.62	½	296.881	7013.81			
83 ins.	260.752	5410.62	95 ins.	298.452	7088.23			
½	262.323	5476.00	½	300.022	7163.04			

TABLE OF PRESSURES ALLOWABLE ON BOILERS.

Diameter of boiler.	Thickness of plates.	45,000 tensile strength. 1-6, 7,500.		50,000 tensile strength. 1-6, 8,333.3.		55,000 tensile strength. 1-6, 9,166.6.		60,000 tensile strength. 1-6, 10,000.		65,000 tensile strength. 1-6, 10,833.3.		70,000 tensile strength. 1-6, 11,666.6.	
		Press. ure.	20 per ct. addi- tional.	Press. ure.	20 per ct. addi- tional.	Press. ure.	20 per ct. addi- tional.	Press. ure.	20 per ct. addi- tional.	Press. ure.	20 per ct. addi- tional.	Press. ure.	20 per ct. addi- tional.
36 inches.	.1875	78.12	93.74	86.8	104.16	95.48	114.57	104.16	124.99	112.84	135.4	121.52	145.82
	.21	87.5	105.	97.21	116.65	106.94	128.3	116.66	139.99	126.38	151.65	136.11	163.33
	.23	95.83	114.99	106.47	127.76	117.12	140.54	127.77	153.32	138.41	166.00	149.07	178.88
	.25	104.16	124.99	115.74	138.88	127.31	152.77	138.88	166.65	150.46	180.55	162.03	194.43
	.26	108.33	129.99	120.37	144.44	132.4	158.88	144.44	173.32	156.48	187.77	168.51	202.21
	.29	120.83	144.99	134.25	161.1	147.68	177.21	161.11	193.33	174.53	209.43	187.90	225.48
	.3125	130.2	156.24	144.67	173.6	159.14	190.96	173.6	208.32	188.07	225.68	202.5	243.04
	.33	137.5	165.	152.77	183.32	168.05	201.66	183.33	219.99	198.61	238.33	213.88	256.65
	.35	145.83	174.99	162.03	194.43	178.23	213.87	194.44	233.32	210.64	252.76	226.84	272.20
	.375	156.25	187.5	173.61	208.33	190.97	229.16	208.33	249.99	225.69	271.82	243.05	291.66
38 inches.	.1875	74.01	88.89	82.23	98.67	90.46	108.54	98.68	118.41	106.9	128.28	115.13	138.16
	.21	82.89	99.46	92.1	110.52	101.31	121.57	110.52	132.62	119.73	143.67	128.93	154.71
	.23	90.78	108.93	100.87	121.04	110.96	133.15	121.05	145.26	131.13	157.35	141.22	169.46
	.25	98.68	118.41	109.64	131.56	120.61	144.73	131.57	157.88	142.54	171.04	153.5	184.20
	.26	102.63	123.15	114.03	136.83	125.43	150.51	136.84	164.2	148.24	177.88	159.64	191.56
	.29	114.47	137.36	127.19	152.62	139.91	167.89	152.63	183.15	165.35	198.42	178.06	213.67
	.3125	123.35	148.02	137.	164.4	150.76	180.91	164.47	197.36	178.17	213.8	191.88	230.25
	.33	130.26	156.31	144.73	173.67	159.2	191.04	173.68	208.41	188.15	225.78	202.62	243.14
	.35	138.15	165.78	153.5	184.21	168.85	202.62	184.21	221.05	199.56	239.47	214.91	257.89
	.375	148.	177.60	164.73	197.67	180.91	217.09	197.36	236.83	213.81	250.57	230.26	276.31
40 inches.	.1875	70.31	84.37	78.12	93.74	85.93	103.11	93.75	112.5	101.56	121.87	109.37	131.24
	.21	78.75	94.50	87.49	104.98	96.24	115.48	105.	126.	113.74	136.48	122.49	146.98
	.23	86.25	103.5	95.83	114.99	105.41	126.49	115.	138.	124.58	149.49	134.16	160.95
	.25	93.75	112.5	104.16	124.99	114.58	137.49	125.	150.	135.41	162.49	145.83	174.99
	.26	97.5	117.	108.33	129.99	119.16	142.99	130.	156.	140.83	168.99	151.66	181.99
	.29	108.75	130.5	120.83	141.99	132.92	159.49	145.	174.	157.08	188.49	169.16	202.99
	.3125	117.18	140.61	130.2	156.24	143.22	171.86	156.25	187.45	169.27	203.12	182.29	218.74
	.33	123.75	148.5	137.49	164.98	151.24	181.48	165.	198.	178.74	214.48	192.49	230.98
	.35	131.25	157.5	145.83	174.99	160.41	192.49	175.	210.	189.58	227.49	204.16	244.99
	.375	140.62	168.74	156.24	187.48	171.87	206.24	187.5	225.	203.12	243.74	218.74	262.48

TABLE OF PRESSURES ALLOWABLE ON BOILERS.—Continued.

Diameter of boiler.	Thickness of plate.	45,000 tensile strength. 1-6, 7,500.		50,000 tensile strength. 1-6, 8,333.3.		55,000 tensile strength. 1-6, 9,166.6.		60,000 tensile strength. 1-6, 10,000.		65,000 tensile strength. 1-6, 10,833.3.		70,000 tensile strength. 1-6, 11,666.6.	
		Press-ure.	20 per ct. addi- tional.	Press-ure.	20 per ct. addi- tional.	Press-ure.	20 per ct. addi- tional.	Press-ure.	20 per ct. addi- tional.	Press-ure.	20 per ct. addi- tional.	Press-ure.	20 per ct. addi- tional.
42 inches.	.1875	66.96	80.35	74.40	89.28	81.84	99.20	89.28	107.13	96.72	116.06	104.16	124.99
	.21	75.-	90.-	83.32	99.99	91.66	109.99	100.-	120.-	108.33	129.99	116.66	139.99
	.23	82.14	98.56	91.23	109.51	100.39	120.46	109.52	131.42	118.65	142.38	127.77	153.32
	.25	89.28	107.13	99.2	119.04	109.12	130.94	119.04	142.84	128.96	154.75	138.88	166.65
	.26	92.85	111.42	103.17	123.8	113.49	136.18	123.8	148.56	134.12	160.94	144.44	173.32
	.29	103.57	124.28	115.07	138.08	126.5	151.85	138.09	165.7	149.6	179.52	161.11	193.33
	.3125	111.6	133.92	124.-	148.8	136.4	163.68	148.74	178.56	161.2	193.44	173.61	208.23
	.33	117.85	141.42	130.94	157.12	144.04	172.84	157.14	188.56	170.23	204.27	183.33	219.99
	.35	125.-	150.-	138.88	166.65	152.77	183.32	166.66	199.99	180.55	216.66	194.44	233.32
	.375	133.92	160.7	148.8	178.56	163.68	196.40	178.57	214.28	193.45	232.14	208.33	249.99
44 inches.	.1875	63.92	76.7	71.02	85.22	78.12	93.73	85.22	102.26	92.32	110.78	99.42	119.3
	.21	71.59	85.9	79.54	95.44	87.49	104.98	95.45	114.64	103.4	124.08	111.36	133.63
	.23	78.4	94.08	87.12	104.54	95.83	114.99	104.54	125.44	113.25	135.9	121.96	146.35
	.25	85.22	102.26	94.69	113.62	104.16	124.99	113.63	136.35	123.1	147.72	131.56	159.07
	.26	88.63	106.35	98.48	118.17	108.33	129.99	118.18	141.81	128.02	153.62	137.87	165.44
	.29	98.86	118.63	109.84	131.80	120.83	144.99	131.81	158.17	142.79	171.33	153.78	184.53
	.3125	106.53	127.83	118.36	142.03	130.2	156.24	142.04	170.44	153.88	184.65	165.71	198.85
	.33	112.5	135.-	124.99	149.98	137.49	164.98	150.-	180.-	162.49	194.98	174.99	209.98
	.35	119.31	143.17	132.57	159.08	145.83	174.99	159.09	190.9	172.34	206.8	185.6	222.72
	.375	127.81	153.37	142.04	170.44	156.24	187.48	170.45	204.54	184.65	221.58	198.86	238.63
46 inches.	.1875	61.14	73.36	67.93	81.51	74.72	89.66	81.51	97.81	88.31	105.97	95.1	114.12
	.21	68.47	82.16	76.08	91.29	83.69	100.42	91.3	109.56	98.91	118.69	106.52	127.82
	.23	75.-	90.-	83.33	100.-	91.66	109.99	100.-	120.-	108.33	129.99	116.66	139.99
	.25	81.51	97.81	90.57	108.68	99.63	119.55	108.69	130.42	117.75	141.3	126.8	152.16
	.26	84.78	101.73	94.2	113.04	103.62	124.34	113.04	135.64	122.46	146.95	131.88	158.25
	.29	94.56	113.47	105.07	126.-	115.57	138.68	126.09	151.3	136.59	163.92	147.1	176.52
	.3125	101.9	122.28	113.21	135.86	124.54	149.44	135.86	163.03	147.19	176.62	158.51	190.21
	.33	107.6	129.12	119.56	143.47	131.52	157.82	143.97	172.16	155.43	186.51	167.39	200.86
	.35	114.13	136.95	126.8	152.16	139.49	167.38	152.17	182.6	164.85	197.82	177.53	213.03
	.375	122.28	146.73	135.86	163.03	149.45	179.34	163.04	195.64	176.62	211.94	190.21	228.25

TABLE OF PRESSURES ALLOWABLE ON BOILERS.—Continued.

Diameter of boiler.	Thickness of plates.	45,000 tensile strength. 1-6, 7,500.		50,000 tensile strength. 1-6, 8,333.3.		55,000 tensile strength. 1-6, 9,166.6.		60,000 tensile strength. 1-6, 10,000.		65,000 tensile strength. 1-6, 10,833.3.		70,000 tensile strength. 1-6, 11,666.6.	
		Press-ure.	20 per ct. additional.	Press-ure.	20 per ct. additional.	Press-ure.	20 per ct. additional.	Press-ure.	20 per ct. additional.	Press-ure.	20 per ct. additional.	Press-ure.	20 per ct. additional.
48 inches.	.1875	58.59	70.30	65.1	78.12	71.61	85.93	78.12	93.74	84.63	101.55	91.13	109.35
	.21	65.62	78.74	72.91	87.49	80.2	96.24	87.49	104.98	94.79	113.74	102.08	122.49
	.23	71.87	86.24	79.85	95.82	87.84	105.4	95.83	114.99	103.81	124.57	111.8	133.16
	.25	78.12	93.74	86.8	104.16	95.48	114.57	104.16	124.99	112.84	135.4	121.52	145.82
	.26	81.25	97.50	90.27	108.32	99.3	119.16	108.33	129.99	117.36	140.83	126.38	151.65
	.29	90.62	108.74	100.69	120.82	110.76	132.91	120.83	144.99	130.9	157.08	140.97	169.16
	.3125	97.65	117.18	108.5	130.2	119.33	143.2	130.21	156.25	141.05	169.26	151.9	182.28
	.33	103.12	123.74	114.58	137.49	126.04	151.24	137.5	165.	148.95	178.74	160.41	192.49
	.35	109.37	131.24	121.52	145.82	133.67	160.4	145.83	174.99	157.98	189.57	170.13	204.15
	.375	117.18	140.61	130.2	156.24	143.22	171.86	156.25	187.50	169.27	203.12	182.29	218.74
54 inches.	.1875	52.08	62.49	57.87	69.44	63.65	76.38	69.44	82.44	75.23	90.27	81.01	97.21
	.21	58.88	69.99	64.81	77.77	71.25	85.54	77.77	93.32	84.25	101.1	90.74	108.88
	.23	63.88	76.65	70.98	85.17	78.08	93.69	85.18	102.21	92.28	110.73	99.38	119.25
	.25	69.44	83.32	77.16	92.59	84.87	101.84	92.59	111.10	100.3	120.36	108.02	129.62
	.26	72.22	86.66	80.24	96.28	88.27	105.92	96.29	115.54	104.31	125.17	112.44	134.8
	.29	80.55	96.66	89.5	107.40	88.45	118.14	107.41	128.88	116.35	139.6	125.3	150.36
	.3125	86.8	104.16	96.44	115.72	106.09	127.30	115.55	138.66	125.38	150.45	135.03	162.03
	.33	91.66	109.99	101.84	122.22	112.03	134.43	122.22	146.66	132.4	158.88	142.59	171.10
	.35	97.22	116.66	108.02	129.62	118.82	142.58	129.62	155.54	140.43	168.51	151.23	181.47
	.375	104.16	124.99	115.74	138.88	127.31	152.77	138.88	166.65	150.46	180.55	162.03	194.43
60 inches.	.1875	46.87	56.24	52.08	62.49	57.29	68.74	62.5	75.	67.7	81.24	72.91	87.49
	.21	52.5	63.	58.33	69.99	64.16	76.99	69.99	84.	75.83	90.99	81.66	97.99
	.23	57.5	69.	63.88	76.65	70.27	84.32	76.66	91.99	83.05	99.66	89.44	107.32
	.25	62.5	75.	69.44	83.32	76.38	91.65	83.33	99.99	90.27	108.32	97.22	116.66
	.26	65.	78.	72.22	86.66	79.44	95.32	86.66	103.99	93.88	112.65	101.11	121.33
	.29	72.5	87.	80.55	96.66	88.61	106.33	96.66	115.99	104.72	125.66	112.77	135.32
	.3125	78.12	93.74	86.8	104.16	95.48	114.57	104.18	124.99	112.95	135.54	121.52	145.82
	.33	82.5	99.	91.66	109.99	100.83	120.99	110.22	132.	119.16	142.99	128.33	153.99
	.35	87.5	105.	97.22	116.66	106.94	128.32	116.66	139.99	126.38	151.65	136.11	163.33
	.375	93.75	112.5	104.16	124.99	114.58	137.49	125.	150.	135.41	162.49	145.83	174.99

TABLE OF PRESSURES ALLOWABLE ON BOILERS.—Continued.

Diameter of Boiler.	Thickness of Plates.	45,000 tensile strength. 1-6, 7,500		50,000 tensile strength. 1-6, 8,333.3		55,000 tensile strength. 1-6, 9,166.6		60,000 tensile strength. 1-6, 10,000.		65,000 tensile strength. 1-6, 10,833.3		70,000 tensile strength. 1-6, 11,666.6	
		Press-ure.	20 perct. addi-tional.	Press-ure.	20 perct. addi-tional.	Press-ure.	20 perct. addi-tional.	Press-ure.	20 perct. addi-tional.	Press-ure.	20 perct. addi-tional.	Press-ure.	20 perct. addi-tional.
66 inches.	.1875	42.61	51.13	47.34	56.8	52.07	62.49	56.81	68.17	61.55	73.86	66.28	79.53
	.21	47.72	57.26	53.	63.63	58.33	69.99	63.63	76.35	68.93	82.71	74.24	83.08
	.23	52.27	62.72	58.	69.69	63.88	76.65	69.09	83.62	75.5	90.6	81.31	97.57
	.25	56.81	68.17	63.13	75.75	69.44	83.32	75.75	90.90	82.07	98.48	88.37	106.04
	.26	59.09	70.9	65.65	78.78	72.22	86.66	78.78	94.53	85.35	102.42	91.91	110.29
	.29	65.90	79.08	73.23	87.87	80.55	96.66	87.87	105.44	95.2	114.24	102.52	123.02
	.3125	71.	85.2	78.91	94.69	86.66	104.16	94.69	113.62	102.58	123.09	110.47	132.56
	.33	75.56	90.	83.33	99.99	91.66	109.99	99.99	120.	108.33	129.99	116.66	139.99
	.35	79.56	95.47	88.38	106.05	97.22	116.66	106.	127.27	114.89	137.86	123.73	148.47
	.375	85.22	102.26	94.69	113.62	104.16	124.99	113.62	136.34	123.1	147.72	132.57	159.08
72 inches.	.1875	39.06	46.87	43.4	52.08	47.74	57.28	52.08	62.49	56.42	67.70	60.76	72.91
	.21	43.75	52.5	48.6	58.33	53.47	64.16	58.33	69.99	63.19	75.82	68.05	81.66
	.23	47.91	57.49	53.24	63.88	58.56	70.27	63.88	76.65	69.21	83.05	74.53	89.43
	.25	52.08	62.49	57.87	69.44	63.65	76.38	69.44	83.32	75.22	90.26	81.01	97.21
	.26	54.16	64.99	60.18	72.21	66.2	79.44	72.22	86.66	78.24	93.88	84.35	101.10
	.29	60.41	72.49	67.12	80.54	73.84	88.60	80.55	96.66	87.26	104.71	93.98	112.77
	.3125	65.10	78.12	72.33	86.8	79.57	95.48	86.8	104.16	94.03	112.83	101.27	121.52
	.33	68.75	82.5	76.38	91.65	84.02	100.82	91.66	109.99	99.3	119.16	106.94	128.32
	.35	72.91	87.49	81.01	97.21	89.11	106.93	97.22	116.66	105.32	126.38	113.42	136.1
	.375	78.12	93.74	86.8	104.16	95.48	114.57	104.16	124.90	112.84	135.43	121.53	145.82
78 inches.	.1875	36.05	43.21	40.06	48.07	44.07	52.87	48.07	57.68	52.08	62.49	56.08	67.29
	.21	40.38	48.45	44.87	53.84	49.35	59.22	53.84	64.60	58.33	69.99	62.82	75.38
	.23	44.23	53.07	49.14	58.96	54.05	64.86	64.4	70.71	63.88	76.65	68.80	82.56
	.25	48.07	57.68	53.41	64.09	58.76	70.5	64.4	76.92	69.44	83.32	74.78	89.73
	.26	50.-	60.-	55.55	66.66	66.11	73.33	66.66	79.99	72.22	86.65	77.77	93.32
	.29	55.76	66.91	61.96	74.35	68.16	81.79	74.35	89.22	80.55	96.66	86.75	104.1
	.3125	60.09	72.1	66.77	80.12	73.45	88.14	80.12	96.14	86.8	104.16	93.48	112.17
	.33	63.46	76.15	70.51	84.61	77.56	93.07	84.61	101.53	91.66	109.99	98.71	118.45
	.35	67.3	80.76	74.78	89.73	82.26	98.71	89.74	107.68	97.22	116.66	104.70	125.64
	.375	72.11	86.53	80.18	96.14	88.14	105.76	96.15	115.38	104.16	124.99	112.17	134.6

TABLE OF PRESSURES ALLOWABLE ON BOILERS.—Continued.

Diameter of Boiler.	Thickness of Plates.	45,000 tensile strength. 1-6, 7,500.		50,000 tensile strength. 1-6, 8,333.3.		55,000 tensile strength. 1-6, 9,166.6.		60,000 tensile strength. 1-6, 10,000.		65,000 tensile strength. 1-6, 10,833.3.		70,000 tensile strength. 1-6, 11,6666.6.	
		Press-ure.	20 perct. addi-tional.	Press-ure.	20 perct. addi-tional.	Press-ure.	20 perct. addi-tional.	Press-ure.	20 perct. addi-tional.	Press-ure.	20 perct. addi-tional.	Press-ure.	20 perct. addi-tional.
84 inches.	.1875	33.48	40.17	37.2	44.68	40.92	49.1	44.64	53.56	48.36	58.03	52.08	62.49
	.21	37.5	45.	41.66	49.99	45.83	54.99	50.	60.	54.16	64.99	58.33	69.99
	.23	41.02	49.22	45.63	54.75	50.19	60.22	54.75	65.71	59.32	71.18	63.65	76.38
	.25	44.64	53.56	49.6	59.52	54.56	65.47	59.52	71.42	64.48	77.37	69.44	83.32
	.26	46.42	55.7	51.58	61.89	56.74	68.08	61.9	74.28	67.05	80.46	72.22	86.66
	.29	51.78	62.13	57.53	69.03	63.29	75.94	69.04	82.84	74.8	89.76	80.55	96.66
	.3125	55.8	66.96	62.	74.4	68.2	81.84	74.4	89.28	80.6	96.72	86.8	104.16
	.33	58.92	70.7	65.47	78.56	72.02	86.42	78.57	94.28	85.11	102.13	91.66	109.99
	.35	62.5	75.	69.44	83.33	76.38	91.65	83.33	99.99	90.27	108.32	97.22	116.66
	.375	66.96	80.35	74.4	89.28	81.84	98.2	89.28	107.13	96.72	116.c6	104.16	124.99
90 inches.	.1875	31.25	37.5	34.72	41.66	38.19	45.82	41.66	49.99	45.13	54.15	48.68	58.33
	.21	35.	42.	38.88	46.65	42.77	51.32	46.66	55.99	50.55	60.66	54.44	65.32
	.23	38.33	45.99	42.59	51.10	46.85	56.22	51.11	61.33	55.37	66.44	59.62	71.54
	.25	41.66	49.99	46.29	55.44	50.92	61.1	55.55	66.66	60.18	72.21	64.81	77.77
	.26	43.33	51.99	48.14	57.76	52.96	63.55	57.77	69.32	62.59	75.1	67.4	80.88
	.29	48.33	57.99	53.7	64.44	59.07	70.8	64.44	77.32	69.81	83.77	75.18	90.21
	.3125	52.08	62.49	57.86	69.43	63.65	76.38	69.44	83.32	75.23	90.27	81.01	97.21
	.33	55.	66.	61.11	73.33	67.22	80.66	73.33	87.99	79.44	95.32	85.55	102.66
	.35	58.33	69.99	64.81	77.77	71.29	85.54	77.77	93.32	84.25	101.1	90.72	108.88
	.375	62.5	75.	69.44	83.32	76.38	91.65	83.33	99.99	90.27	108.32	97.22	116.66
96 inches.	.1875	29.29	35.14	32.55	39.06	35.8	42.96	39.06	46.87	42.31	50.77	45.57	54.68
	.21	32.81	39.37	36.45	43.74	40.1	48.12	43.75	52.5	47.39	56.86	51.04	61.24
	.23	35.93	43.11	39.93	47.91	43.92	52.7	47.91	57.49	51.9	62.28	55.9	67.08
	.25	39.06	46.87	43.4	52.08	47.74	57.28	52.08	62.49	56.42	67.67	60.76	72.91
	.26	40.62	48.74	45.14	54.16	49.65	59.58	54.16	64.99	58.78	70.53	63.19	75.82
	.29	45.31	54.37	50.34	60.4	55.38	66.45	60.41	72.49	65.45	78.54	70.48	84.57
	.3125	48.82	58.58	54.25	65.1	59.67	71.6	65.1	78.12	70.52	84.62	75.95	91.14
	.33	51.56	61.87	57.29	68.74	63.02	75.62	68.75	82.5	74.47	89.36	80.2	96.24
	.35	54.68	65.61	60.76	72.91	66.83	80.19	72.91	87.49	78.99	94.78	85.06	102.07
	.375	58.58	70.29	65.1	78.12	71.61	85.93	78.12	93.74	84.63	101.55	91.14	109.6

PRESSURE ALLOWABLE ON BOILERS.

TABLE OF PRESSURES ALLOWED UNDER THE PROVISIONS OF THE SPECIAL ACT OF CONGRESS RELATING TO THE LIMITATION OF STEAM-PRESSURE OF VESSELS USED EXCLUSIVELY FOR TOWING AND CARRYING FREIGHT ON THE MISSISSIPPI RIVER AND ITS TRIBUTARIES, APPROVED JANUARY 6, 1874.

Thickness of iron.	34 ins. diam.	36 ins. diam.	38 ins. diam.	40 ins. diam.	42 ins. diam.	44 ins. diam.	46 ins. diam.	48 ins. diam.	50 ins. diam.	52 ins. diam.	54 ins. diam.	56 ins. diam.	58 ins. diam.	60 ins. diam.
Inches.	Pds.	Pds.	Pds.	Pds.	Pds.	Pds.	Pds.	Pds.	Pds.	Pds.	Pds.	Pds.	Pds.	Pds.
.19	140.82	133.	126.	119.70	114.	108.81	104.08	99.75	95.75	92.07	88.66	85.50	82.55	79.80
.20	148.23	140.	132.63	126.	120.	114.54	109.56	105.	100.80	96.92	93.33	90.	85.90	84.
.21	155.64	147.	139.26	132.30	126.	120.27	115.04	110.25	105.84	101.77	98.	94.50	91.24	88.20
.22	163.05	154.	149.50	138.00	132.	126.	120.52	115.50	111.08	106.81	102.66	99.	95.69	92.40
.23	170.47	161.	152.52	144.90	138.	131.72	126.	120.75	115.92	111.46	107.33	103.50	99.93	96.60
.24	177.88	168.	159.15	151.20	144.	137.45	131.47	126.	120.95	117.07	112.	108.	104.27	100.80
.25	185.29	175.	165.79	157.50	150.	143.28	136.95	131.25	126.	121.15	116.66	112.50	108.62	105.
.26	192.70	182.	172.42	163.80	156.	148.80	142.43	136.50	131.04	126.	121.33	117.	113.	109.20
.27	200.11	189.	179.05	170.10	162.	154.63	147.91	141.75	136.08	130.84	126.	121.50	117.31	113.40
.28	207.53	191.	185.68	176.40	168.	160.36	152.52	147.	141.11	135.69	130.66	126.	121.65	117.60
.29	214.94	203.	192.31	182.70	174.	166.04	158.87	152.25	146.16	140.53	134.25	130.50	126.	121.80
.30	222.35	210.	198.94	189.	180.	171.81	164.34	157.50	151.50	145.38	140.	135.	130.34	126.
.31	229.76	217.	205.57	195.30	186.	177.54	169.82	162.75	156.24	150.23	144.66	139.50	134.68	130.

The above table gives the steam-pressure allowed on boilers used on freight and towing steamers, the standard pressure being 150 pounds for a boiler 42 inches diameter and .25 of an inch thick. To find the pressure required on other size boilers (not given in the above table), multiply 12,600 by the thickness and .25 of an inch thick. To find the pressure required on other size boilers (not given in the above table), multiply 12,600 by the thickness and divide by the radius, or half the diameter.

The U. S. rule for finding the dimension of boilers is as follows: RULE.—Multiply one-sixth ($\frac{1}{6}$) of the lowest tensile strength found stamped on any plate in the cylindrical shell by the thickness (expressed in inches or parts of an inch) of the thinnest plate in the same cylindrical shell, and divide by the radius, or half diameter (also expressed in inches), and the quotient will be the pressure allowable per square inch of surface for single riveting, to which add twenty per cent for double riveting, etc.

SUPERFICIAL AREA OF TUBES IN SQUARE FEET.

LENGTH OF TUBES IN FEET.

Diam in Inches	1	2	3	4	5	6	7	8	9	10	11	12	13	14	15	16
½	.1309	.2618	.3927	.5236	.6545	.7854	.9163	1.0472	1.1781	1.309	1.4399	1.5708	1.7017	1.9326	1.9635	2.0944
⅝	.1636	.3272	.4909	.6545	.8181	.9817	1.1454	1.309	1.4726	1.6362	1.7999	1.9635	2.1271	2.2907	2.4544	2.618
¾	.1797	.3594	.539	.7187	.8984	1.0781	1.2578	1.4375	1.6171	1.7978	1.9775	2.1572	2.3369	2.5155	2.6952	2.8749
1	.2618	.5236	.7854	1.0472	1.309	1.5708	1.8326	2.0944	2.3562	2.618	2.8978	3.1416	3.4034	3.6652	3.927	4.1888
1¼	.3272	.6545	.9817	1.309	1.6362	1.9635	2.2907	2.618	2.9451	3.2725	3.5997	3.927	4.2542	4.5815	4.9087	5.236
1½	.3927	.7854	1.1781	1.5708	1.9635	2.3562	2.7489	3.1416	3.5343	3.927	4.3197	4.7124	5.1051	5.4978	5.8905	6.2832
1¾	.4581	.9163	1.3744	1.8326	2.2907	2.7489	3.207	3.6652	4.1233	4.5815	5.0396	5.4978	5.9559	6.4141	6.8722	7.3304
2	.5236	1.0472	1.5708	2.0944	2.618	3.1416	3.6652	4.1888	4.7124	5.236	5.7596	6.2832	6.8068	7.3304	7.854	8.3776
2½	.6545	1.309	1.9635	2.618	3.2725	3.927	4.5815	5.236	5.8905	6.545	7.1995	7.854	8.5085	9.163	9.8175	10.472
3	.7854	1.5708	2.3562	3.1416	3.927	4.7124	5.4978	6.2832	7.0686	7.854	8.6394	9.4248	10.2102	10.9956	11.781	12.5664
3½	.9163	1.8326	2.7489	3.6652	4.5815	5.4978	6.4141	7.3304	8.2467	9.163	10.0793	10.9956	11.9119	12.8282	13.7445	14.6608
4	1.0472	2.0944	3.1416	4.1888	5.236	6.2832	7.3304	8.3776	9.4248	10.472	11.5192	12.5664	13.6136	14.6608	15.708	16.7552
4½	1.1781	2.3562	3.5343	4.7124	5.8905	7.0686	8.2467	9.4248	10.6029	11.781	12.9591	14.1372	15.3153	16.4934	17.6715	18.8496
5	1.309	2.618	3.927	5.236	6.545	7.854	9.163	10.472	11.781	13.09	14.399	15.708	17.017	18.326	19.635	20.944
6	1.5708	3.1416	4.7124	6.2832	7.854	9.4248	10.9956	12.5664	14.1372	15.708	17.2788	18.8496	20.4204	21.9912	23.562	25.1328
7	1.8326	3.6652	5.4978	7.3304	9.163	10.9956	12.8282	14.6608	16.4934	18.326	20.1586	21.9912	23.8338	25.6564	27.489	29.3216

SUPERFICIAL AREA OF BOILER FLUES IN SQUARE FEET.

Length of Flues in Feet.

Diameter in Inches.	1	2	3	4	5	6	7	8	9	10	11	12	14	15	16	18	20	24	30
8	2.094	4.189	6.283	8.378	10.472	12.566	14.661	16.755	18.849	20.944	23.038	25.133	29.3216	1.416	33.51	37.699	41.888	48.171	62.83
9	2.356	4.712	7.068	9.425	11.781	14.137	16.493	18.849	21.206	23.562	25.918	28.274	32.987	35.343	37.699	42.412	47.124	56.549	70.68
10	2.618	5.236	7.854	10.472	13.09	15.708	18.326	20.944	23.562	26.18	28.798	31.416	36.652	39.27	41.888	47.124	52.36	62.832	78.54
11	2.879	5.759	8.639	11.519	14.399	17.279	20.158	23.038	25.918	28.798	31.678	34.557	40.317	43.197	46.077	51.835	57.596	69.115	86.394
12	3.1416	6.283	9.424	12.566	15.708	18.849	21.991	25.133	28.274	31.416	34.557	37.699	43.982	47.124	50.266	56.549	62.832	75.398	94.248
13	3.403	6.806	10.21	13.613	17.017	20.42	23.824	27.227	30.6306	34.034	37.437	40.841	47.647	51.051	54.454	61.261	68.068	81.681	102.102
14	3.665	7.33	10.995	14.66	18.326	21.991	25.656	29.321	32.987	36.652	40.317	43.982	51.313	54.978	58.643	65.973	73.304	86.965	109.956
15	3.927	7.854	11.781	15.708	19.635	23.562	27.489	31.416	35.343	39.27	43.197	47.124	54.978	58.905	62.832	70.686	78.54	94.248	117.81
16	4.189	8.377	12.566	16.755	20.944	25.133	29.3216	33.5104	37.699	41.88	46.077	50.265	58.643	62.832	67.021	75.398	83.775	100.53	125.664
17	4.450	8.901	13.352	17.802	22.253	26.704	31.154	35.605	40.055	44.506	48.956	53.407	62.308	66.759	71.209	80.111	89.012	106.814	133.518
18	4.712	9.424	14.137	18.849	23.562	28.274	32.987	37.699	42.412	47.12	51.836	56.549	65.973	70.686	75.398	84.823	94.248	113.097	141.377
20	5.236	10.472	15.708	20.944	26.18	31.416	36.652	41.888	47.124	52.36	57.596	62.832	73.304	78.54	83.776	94.248	104.72	125.664	157.08
24	6.283	13.09	18.849	25.087	31.416	37.699	43.992	50.175	56.549	62.832	69.115	75.398	87.965	94.248	100.531	113.097	125.664	150.807	188.496
30	7.854	15.708	23.562	31.416	39.27	47.124	54.978	62.832	70.686	78.54	86.394	94.248	109.956	117.81	125.664	141.377	157.08	187.496	235.62
36	9.1248	18.8496	28.274	37.699	47.124	56.549	65.973	75.398	84.823	94.248	103.673	113.097	131.947	141.372	150.796	169.64	188.496	226.194	282.744
42	10.9956	21.991	32.987	43.982	54.978	65.973	76.969	87.965	98.96	109.956	120.951	131.947	153.938	164.934	175.929	197.921	219.912	263.894	329.868

SUPERFICIAL AREA OF HALF THE SHELL OF A CYLINDRICAL BOILER.

LENGTH OF SHELL IN FEET.

Diameter in Inches.	1	10	11	12	13	14	15	16	17	18	19	20	22	24	30
30	4.43	44.32	48.752	53.184	57.616	62.048	66.48	70.912	75.344	79.776	84.108	88.64	97.504	105.368	132.96
36	4.71	47.128	51.841	55.554	60.266	64.979	69.692	74.405	79.118	83.83	88.533	94.256	103.682	113.107	149.384
42	5.498	54.978	60.476	65.974	71.471	76.969	82.467	87.965	93.463	98.96	104.458	109.956	120.952	131.947	164.934
45	5.89	58.905	64.795	70.686	76.576	82.467	88.363	94.26	100.156	106.053	111.949	117.846	129.591	141.372	176.726
48	6.28	62.832	69.115	75.398	81.682	87.965	94.248	100.531	106.814	113.098	119.381	125.664	138.23	150.797	188.496
54	7.07	70.686	77.755	84.823	91.892	98.96	106.029	113.098	120.166	127.235	134.303	141.372	155.509	169.646	212.058
60	7.85	78.54	86.394	94.248	102.102	109.956	117.81	125.664	133.518	141.372	149.226	157.08	172.788	188.496	235.62
66	8.64	86.394	95.033	103.673	112.312	120.952	129.591	138.231	146.87	155.509	164.149	172.788	189.017	206.336	259.182
72	9.425	94.248	103.673	113.098	122.522	131.947	141.372	150.797	160.222	169.646	179.071	188.496	207.346	226.195	282.744
78	10.21	102.102	112.312	122.522	132.733	142.943	153.153	163.363	173.573	183.784	193.994	204.204	224.424	245.045	306.306
84	10.995	109.956	120.952	131.947	142.943	153.939	164.934	175.930	186.925	197.921	208.916	219.912	241.903	263.894	329.868

GUIDE POSTS ON THE ENGINEER'S JOURNEY. 157

LAP-WELDED BOILER TUBES.

External Diameter.	Thickness.	Internal Area.	Weight per foot.	External Diameter.	Thickness.	Internal Area.	Weight per foot.
Ins.	Ins.	Sq. Ins.	Lbs.	Ins.	Ins.	Sq. Ins.	Lbs.
1	.072	.057	.70	6	.165	25.50	10.16
1¼	.072	.096	.90	7	.165	34.80	11.90
1½	.083	1.39	1.24	8	.165	45.79	13.65
2	.095	2.55	1.91	10	.203	71.97	21.00
2½	.109	4.09	2.75	12	.229	103.74	28.50
3	.109	6.08	3.33	14	.248	143.18	36.00
3½	.120	8.35	4.28	15	.259	164.71	40.60
4	.134	10.99	5.47	16	.270	187.66	45.20
4½	.134	14.12	6.17	18	.292	238.22	54.81
5	.148	17.49	7.58	20	.320	294.37	66.76

WEIGHT OF 100 RIVETS IN POUNDS. (TRAUTWINE).

Length of Shank in Inches.	Diameter of Rivets in Inches.							
	⅜	½	⅝	¾	⅞	1	1⅛	1¼
½	3.0	8.5	....	....	....	....	....	....
¾	3.8	9.9	17.3	....	....	....	....	....
1	4.6	11.2	19.4	25.6	38.9	....	....	....
¼	5.4	12.6	21.5	28.7	43.1	65.3	91.5	123.
½	6.2	13.9	23.7	31.8	47.3	70.7	98.4	133.
¾	6.9	15.3	25.8	34.9	51.4	76.2	105.	142.
2	7.7	16.6	27.9	37.9	55.6	81.6	112.	150.
¼	8.5	18.0	30.0	41.0	59.8	87.1	119.	159.
½	9.2	19.4	32.2	44.1	64.0	92.5	126.	167.
¾	10.0	20.7	34.3	47.1	68.1	98.0	133	176.
3	10.8	22.1	36.4	50.2	72.3	103.	140.	184.
¼	11.5	23.5	38.6	53.3	76.5	109.	147.	193.
½	12.3	24.8	40 7	56.4	80.7	114.	154.	201.
¾	13.1	26.2	42.8	59.4	84.8	120.	161.	210.
4	13.8	27.5	45.0	62.5	89.0	125.	167.	18.
¼	14.6	28.9	47.1	65.6	93.2	131.	174.	27.
½	15.4	30.3	49.2	68.6	97.4	136.	181.	36.
¾	16.2	31.6	51.4	71.7	102.	142.	188.	44.
5	16.9	33.0	53.5	74.8	106.	147.	195.	53.
¼	17.7	34.4	55.6	77.8	110.	153.	202.	61.
½	18.4	35.7	57.7	80.9	114.	158.	209.	270.
¾	19.2	37.1	59.9	84.0	118.	163.	216.	278.
6	20.0	38.5	62.0	87.0	122.	169.	223.	287.
½	21.5	41.2	66.3	93.2	131.	180.	236.	304.
7	23.0	43.9	70.5	99.3	139.	191.	250.	321.
½	24.6	46.6	74.8	106.	147.	202.	264.	338.
8	26.1	49.4	79.0	112.	156.	213.	278.	355.
9	29.2	54.8	87.6	124.	173.	234.	306.	389.
10	32.2	60.3	96.1	136.	189.	256.	333.	423.
11	35.3	65.7	105.	148.	206.	278.	361.	457.
12	38.4	71.2	113.	161.	223.	300.	388.	491.

PROPERTIES OF SATURATED STEAM.

Total pressure per square inch, measured from a vacuum.	Pressure above the atmosphere.	Sensible temperature in Fahrenheit degrees.	Total heat in degrees from zero of Fahrenheit.	Weight of 1 cubic foot of steam.	Relative volume of the steam compared with the water from which it was raised.
Lbs.	Lbs.	Degrees.	Degrees.	Lbs.	
1	....	102.1	1144.5	.0030	20582
2	....	126.3	1151.7	.0058	10721
3	....	141.6	1156.6	.0085	7322
4	....	153.1	1160.1	.0112	5583
5	....	162.3	1162.9	.0138	4527
6	....	170.2	1165.3	.0163	3813
7	....	176.9	1167.3	.0189	3298
8	....	182.9	1169.2	.0214	2909
9	....	188.3	1170.8	.0239	2604
10	....	193.3	1172.3	.0264	2358
11	....	197.8	1173.7	.0289	2157
12	....	202.0	1175.0	.0314	1986
13	....	205.9	1176.2	.0338	1842
14	....	209.6	1177.3	.0362	1720
14.7	....	212.0	1178.1	.0380	1642
15	.3	213.1	1178.4	.0387	1610
16	1.3	216.3	1179.4	.0411	1515
17	2.3	219.6	1180.3	.0435	1431
18	3.3	222.4	1181.2	.0459	1357
19	4.3	225.3	1182.1	.0483	1290
20	5.3	228.0	1182.9	.0507	1229
21	6.3	230.6	1183.7	.0531	1174
22	7.3	233.1	1184.5	.0555	1123
23	8.3	235.5	1185.2	.0580	1075
24	9.3	237.8	1185.9	.0601	1036

PROPERTIES OF SATURATED STEAM.—Continued.

Total pressure per square inch, measured from a vacuum.	Pressure above the atmosphere.	Sensible temperature in Fahrenheit degrees.	Total heat in degrees from zero of Fahrenheit.	Weight of 1 cubic foot of steam.	Relative volume of the steam compared with the water from which it was raised
Lbs.	Lbs.	Degrees.	Degrees.	Lbs.	
25	10.3	240.1	1186.6	.0625	996
26	11.3	242.3	1187.3	.0650	958
27	12.3	244.4	1187.8	.0673	926
28	13.3	246.4	1188.4	.0696	895
29	14.3	248.4	1189.1	.0719	866
30	15.3	250.4	1189.8	.0743	838
31	16.3	252.2	1190.4	.0766	813
32	17.3	254.1	1190.9	.0789	789
33	18.3	255.9	1191.5	.0812	767
34	19.3	257.6	1192.0	.0835	746
35	20.3	259.3	1192.5	.0858	726
36	21.3	260.9	1193.0	.0881	707
37	22.3	262.6	1193.5	.0905	688
38	23.3	264.2	1194.0	.0929	671
39	24.3	265.8	1194.5	.0952	655
40	25.3	267.3	1194.9	.0974	640
41	26.3	268.7	1195.4	.0996	625
42	27.3	270.2	1195.8	.1020	611
43	28.3	271.6	1196.2	.1042	598
44	29.3	273.0	1196.6	.1065	585
45	30.8	274.4	1197.1	.1089	572
46	31.3	275.8	1197.5	.1111	561
47	32.3	277.1	1197.9	.1133	550
48	33.3	278.4	1198.3	.1156	539
49	34.3	279.7	1198.7	.1179	529

PROPERTIES OF SATURATED STEAM.—Continued.

Total pressure per square inch, measured from a vacuum.	Pressure above the atmosphere.	Sensible temperature in Fahrenheit degrees.	Total heat in degree from zero of Fahrenheit	Weight of 1 cubic foot of steam.	Relative volume of the steam compared with the water from which it was raised.
Lbs.	Lbs.	Degrees.	Degrees.	Lbs.	
50	35.3	281.0	1199.1	.1202	518
51	36.3	282.3	1199.5	.1224	509
52	37.3	283.5	1199.9	.1246	500
53	38.3	284.7	1200.3	.1269	491
54	39.3	285.9	1200.6	.1291	482
55	40.3	287.1	1201.0	.1314	474
56	41.3	288.2	1201.3	.1336	466
57	42.3	289.3	1201.7	.1364	458
58	43.3	290.4	1202.0	.1380	451
59	44.3	291.6	1202.4	.1403	444
60	45.3	292.7	1202.7	.1425	437
61	46.3	293.8	1203.1	.1447	430
62	47.3	294.8	1203.4	.1469	424
63	48.3	295.9	1203.7	.1493	417
64	49.3	296.9	1204.0	.1516	411
65	50.3	298.0	1204.3	.1538	405
66	51.3	299.0	1204.6	.1560	399
67	52.3	300.0	1204.9	.1583	393
68	53.3	300.9	1205.2	.1605	388
69	54.3	301.9	1205.5	.1627	383
70	55.3	302.9	1205.8	.1648	378
71	56.3	303.9	1206.1	.1670	373
72	57.3	304.8	1206.3	.1692	368
73	58.3	305.7	1206.6	.1714	363
74	59.3	306.6	1206.9	.1736	359

PROPERTIES OF SATURATED STEAM.—Continued.

Total pressure per square inch, measured from a vacuum.	Pressure above the atmosphere.	Sensible temperature in Fahrenheit degrees.	Total heat in degrees from zero of Fahrenheit.	Weight of 1 cubic foot of steam.	Relative volume of the steam compared with the water from which it was raised.
Lbs.	Lbs.	Degrees.	Degrees.	Lbs.	
75	60.3	307.5	1207.2	.1759	353
76	61.3	308.4	1207.4	.1782	349
77	62.3	309.3	1207.7	.1804	345
78	63.3	310.2	1208.0	.1826	341
79	64.3	311.1	1208.3	.1848	337
80	65.3	312.0	1208.5	.1869	333
81	66.3	312.8	1208.8	.1891	329
82	67.3	313.6	1209.1	.1913	325
83	68.3	314.5	1209.4	.1935	321
84	69.3	315.3	1209.6	.1957	318
85	70.3	316.2	1209.9	.1980	314
86	71.3	317.8	1210.1	.2002	311
87	72.3	318.6	1210.4	.2024	308
88	73.3	319.4	1210.6	.2044	305
89	74.3	320.2	1210.9	.2067	301
90	75.3	321.0	1211.1	.2089	298
91	76.3	321.7	1211.3	.2111	295
92	77.3	322.5	1211.5	.2133	292
93	78.3	323.3	1211.8	.2155	289
94	79.3	324.1	1212.0	.2176	286
95	80.3	324.8	1212.3	.2198	283
96	81.3	325.6	1212.5	.2219	281
97	82.3	326.3	1212.8	.2241	278
98	83.3	327.1	1213.0	.2263	275
99	84.3	327.9	1213.2	.2285	272

PROPERTIES OF SATURATED STEAM — Continued.

Total pressure per square inch, measured from a vacuum.	Pressure above the atmosphere.	Sensible temperature in Fahrenheit degrees.	Total heat in degrees from zero of Fahrenheit.	Weight of 1 cubic foot of steam.	Relative volume of the steam compared with the water from which it was raised.
Lbs.	Lbs.	Degrees.	Degrees.	Lbs.	
100	85.3	327.9	1213.4	.2307	270
101	86.3	328.5	1213.6	.2329	267
102	87.3	329.1	1213.8	.2351	265
103	88.3	329.9	1214.0	.2373	262
104	89.3	330.6	1214.2	.2393	260
105	90.3	331.3	1214.4	.2414	257
106	91.3	331.9	1214.6	.2435	255
107	92.3	332.6	1214.8	.2456	253
108	93.3	333.3	1215.0	.2477	251
109	94.3	334.0	1215.3	.2499	249
110	95.3	334.6	1215.5	.2521	247
111	96.3	335.3	1215.7	.2543	245
112	97.3	336.0	1215.9	.2564	243
113	98.3	336.7	1216.1	.2586	241
114	99.3	337.4	1216.3	.2607	239
115	100.3	338.0	1216.5	.2628	237
116	101.3	338.6	1216.7	.2649	235
117	102.3	339.3	1216.9	.2674	233
118	103.3	339.9	1217.1	.2696	231
119	104.3	340.5	1217.3	.2738	229
120	105.3	341.1	1217.4	.2759	227
121	106.3	341.8	1217.6	.2780	225
122	107.3	342.4	1217.8	.2801	224
123	108.3	343.0	1218.0	.2822	222
124	109.3	343.6	1218.2	.2845	221

PROPERTIES OF SATURATED STEAM.—Continued

Total pressure per square inch, measured from a vacuum.	Pressure above the atmosphere.	Sensible temperature in Fahrenheit degrees.	Total heat in degrees from zero of Fahrenheit.	Weight of 1 cubic foot of steam.	Relative volume of the steam compared with the water from which it was raised.
Lbs.	Lbs.	Degrees.	Degrees.	Lbs.	
125	110.3	344.2	1218.4	.2867	219
126	111.3	344.8	1218.6	.2889	217
127	112.3	345.4	1218.8	.2911	215
128	113.3	346.0	1218.9	.2933	214
129	114.3	346.6	1219.1	.2955	212
130	115.3	347.2	1219.3	.2977	211
131	116.3	347.8	1219.5	.2999	209
132	117.3	348.3	1219.6	.3020	208
133	118.3	348.9	1219.8	.3040	206
134	119.3	349.5	1220.0	.3060	205
135	120.3	350.1	1220.2	.3080	203
136	121.3	350.6	1220.3	.3101	202
137	122.3	351.2	1220.5	.3121	200
138	123.3	351.8	1220.7	.3142	199
139	124.3	352.4	1220.9	.3162	198
140	125.3	352.9	1221.0	.3184	197
141	126.3	353.5	1221.2	.3206	195
142	127.3	354.0	1221.4	.3228	194
143	128.3	354.5	1221.6	.3250	193
144	129.3	355.0	1221.7	.3273	192
145	130.3	355.6	1221.9	.3294	190
146	131.3	356.1	1222.0	.3315	189
147	132.3	356.7	1222.2	.3336	188
148	133.3	357.2	1222.3	.3357	187
149	134.3	357.8	1222.5	.3377	186

PROPERTIES OF SATURATED STEAM.—Continued.

Total pressure per square inch measured from a vacuum.	Pressure above the atmosphere.	Sensible temperature in Fahrenheit degrees.	Total heat in degrees from zero to Fahrenheit.	Weight of 1 cubic foot of steam.	Relative volume of the steam compared with the water from which it is raised.
Lbs.	Lbs.	Degrees.	Degrees.	Lbs.	
150	135.3	358.3	1222.7	.3397	184
155	140.3	361.0	1223.5	.3500	179
160	145.3	363.4	1224.2	.3607	174
165	150.3	366.0	1224.9	.3714	169
170	155.9	368.2	1225 7	.3821	164
175	160.3	370.8	1226.4	.3928	159
180	165.3	372.9	1227.1	.4035	155
185	170.3	375.3	1227.8	.4142	151
190	175.3	377.5	1228.5	.4250	148
195	180.3	379.7	1229.2	.4357	144
200	185.3	381.7	1229 8	.4464	141
210	195.3	386.0	1231.1	.4668	135
220	205 3	389.9	1232.3	.4872	129
230	215.3	393 8	1233.5	.5072	123
240	225 3	397.5	1234.6	.5270	119
250	235.3	401.1	1235.7	.5471	114
260	245.3	404.5	1236.8	.5670	110
270	255.3	407.9	1237.8	.5871	106
280	265.3	411.2	1238.8	.6070	102
290	275.3	414.4	1239.8	.6268	99
300	285 3	417.5	1240.7	.6469	96

FOREIGN TERMS FOR HORSE-POWER, AND VALUES IN UNITS PER MINUTE.

Country.	Terms.	Values per minute.	English values per minute
England	Horse-power	33,000 foot-pounds	33,000 foot-pounds.
France	Force de cheval, or cheval-vapeur	45,000 kilogramme-ters	32,548.2 foot-pounds.
Germany	Pferde-kraft	30,780 fuss-pfunde	34,935 foot-pounds.
Sweden	Hast-kraft	36,000 Skalpund-fot	32,523.6 foot-pounds.
Russia	Syl-lochad	33,000 Fyt-funt	33,000 foot-pounds.

NOTE.—A Pferde-stärke = 480 Fuss-pfunde per second = 24,800 foot-pounds per minute.

VELOCITY OF STEAM ESCAPING INTO THE ATMOSPHERE.

Pressure above atmosphere.	Velocity per second.	Pressure above atmosphere.	Velocity per second.
Pounds.	Feet.	Pounds.	Feet.
1	540	50	1736
2	698	60	1777
3	814	70	1810
4	905	80	1835
5	981	90	1857
10	1232	100	1875
20	1476	110	1889
30	1601	120	1900
40	1681	130	1909

HYPERBOLIC LOGARITHMS.

N.	Log.	N.	Log.	N.	Log.
1.01	0.00995	1.31	0.27003	1.61	0.47623
1.02	0.01980	1.32	0.27763	1.62	0.48243
1.03	0.02156	1.33	0.28518	1.63	0.48858
1.04	0.03922	1.34	0.29267	1.64	0.49470
1.05	0.04879	1.35	0.30010	1.65	0.50077
1.06	0.05827	1.36	0.30748	1.66	0.50682
1.07	0.06766	1.37	0.31481	1.67	0.51282
1.08	0.07696	1.38	0.32208	1.68	0.51879
1.09	0.08718	1.39	0.32930	1.69	0.52473
1.10	0.09531	1.40	0.33647	1.70	0.53063
1.11	0.10436	1.41	0.34359	1.71	0.53649
1.12	0.11333	1.42	0.35066	1.72	0.54232
1.13	0.12222	1.43	0.35767	1.73	0.54812
1.14	0.13103	1.44	0.36464	1.74	0.55389
1.15	0.13976	1.45	0.37156	1.75	0.55962
1.16	0.14842	1.46	0.37844	1.76	0.56531
1.17	0.15700	1.47	0.38526	1.77	0.57098
1.18	0.16551	1.48	0.39204	1.78	0.57661
1.19	0.17395	1.49	0.39878	1.79	0.58222
1.20	0.18232	1.50	0.40547	1.80	0.58779
1.21	0.19062	1.51	0.41211	1.81	0.59313
1.22	0.19885	1.52	0.41871	1.82	0.59884
1.23	0.20701	1.53	0.42527	1.83	0.60432
1.24	0.21511	1.54	0.43178	1.84	0.60977
1.25	0.22314	1.55	0.43825	1.85	0.61519
1.26	0.23111	1.56	0.44469	1.86	0.62058
1.27	0.23902	1.57	0.45108	1.87	0.62584
1.28	0.24686	1.58	0.45742	1.88	0.63127
1.29	0.25464	1.59	0.46373	1.89	0.63658
1.30	0.26236	1.60	0.47000	1.90	0.64185

HYPERBOLIC LOGARITHMS.—Continued.

N.	Log.	N.	Log.	N.	Log.
1.91	0.64710	2.21	0.79299	2.51	0.92028
1.92	0.65233	2.22	0.79751	2.52	0.92426
1.93	0.65752	2.23	0.80200	2.53	0.92822
1.94	0.66269	2.24	0.80648	2.54	0.93216
1.95	0.66783	2.25	0.81093	2.55	0.93609
1.96	0.67294	2.26	0.81536	2.56	0.94001
1.97	0.67803	2.27	0.81978	2.57	0.94301
1.98	0.68310	2.28	0.82418	2.58	0.94779
1.99	0.68813	2.29	0.82855	2.59	0.95166
2.00	0.69315	2.30	0.83291	2.60	0.95551
2.01	0.69813	2.31	0.83725	2.61	0.95935
2.02	0.70310	2.32	0.84157	2.62	0.96317
2.03	0.70804	2.33	0.84587	2.63	0.96698
2.04	0.71295	2.34	0.85015	2.64	0.97078
2.05	0.71784	2.35	0.85442	2.65	0.97456
2.06	0.72271	2.36	0.85866	2.66	0.97833
2.07	0.72755	2.37	0.86289	2.67	0.98208
2.08	0.73237	2.38	0.86710	2.68	0.98582
2.09	0.73716	2.39	0.87129	2.69	0.98954
2.10	0.74194	2.40	0.87547	2.70	0.99325
2.11	0.74669	2.41	0.87963	2.71	0.99695
2.12	0.75142	2.42	0.88377	2.72	1.00063
2.13	0.75612	2.43	0.88789	2.73	1.00430
2.14	0.76081	2.44	0.89200	2.74	1.00796
2.15	0.76547	2.45	0.89609	2.75	1.01160
2.16	0.77011	2.46	0.90010	2.76	1.01523
2.17	0.77471	2.47	0.90422	2.77	1.01885
2.18	0.77932	2.48	0.90826	2.78	1.02245
2.19	0.78391	2.49	0.91228	2.79	1.02604
2.20	0.78846	2.50	0.91629	2.80	1.02962

GUIDE POSTS ON THE ENGINEER'S JOURNEY. 169

HYPERBOLIC LOGARITHMS.—Continued.

N.	Log.	N.	Log.	N.	Log.
2.81	1.03318	3.11	1.13462	3.41	1.22671
2.82	1.03674	3.12	1.13783	3.42	1.22964
2.83	1.04028	3.13	1.14103	3.43	1.23256
2.84	1.04380	3.14	1.14422	3.44	1.23547
2.85	1.04732	3.15	1.14740	3.45	1.23837
2.86	1.05082	3.16	1.15057	3.46	1.24187
2.87	1.05431	3.17	1.15373	3.47	1.24415
2.88	1.05779	3.18	1.15688	3.48	1.24703
2.89	1.06126	3.19	1.16002	3.49	1.24990
2.90	1.06471	3.20	1.16315	3.50	1.25276
2.91	1.06815	3.21	1.16627	3.51	1.25562
2.92	1.07158	3.22	1.16938	3.52	1.25846
2.93	1.07500	3.23	1.17248	3.53	1.26130
2.94	1.07841	3.24	1.17557	3.54	1.26413
2.95	1.08181	3.25	1.17865	3.55	1.26695
2.96	1.08519	3.26	1.18173	3.56	1.26976
2.97	1.08856	3.27	1.18479	3.57	1.27257
2.98	1.09192	3.28	1.18784	3.58	1.27536
2.99	1.09527	3.29	1.19089	3.59	1.27815
3.00	1.09861	3.30	1.19392	3.60	1.28093
3.01	1.10194	3.31	1.19695	3.61	1.28371
3.02	1.10526	3.32	1.19996	3.62	1.28647
3.03	1.10856	3.33	1.20297	3.63	1.28923
3.04	1.11186	3.34	1.20597	3.64	1.29198
3.05	1.11514	3.35	1.20896	3.65	1.29473
3.06	1.11841	3.36	1.21194	3.66	1.29746
3.07	1.12168	3.37	1.21491	3.67	1.30019
3.08	1.12493	3.38	1.21788	3.68	1.30291
3.09	1.12817	3.39	1.22083	3.69	1.30563
3.10	1.13410	3.40	1.22378	3.70	1.30833

HYPERBOLIC LOGARITHMS.—Continued.

N.	Log.	N.	Log.	N.	Log.
3.71	1.31103	4.01	1.38879	4.31	1.46094
3.72	1.31372	4.02	1.39128	4.32	1.46326
3.73	1.31641	4.03	1.39377	4.33	1.46557
3.74	1.31909	4.04	1.39624	4.34	1.46787
3.75	1.32176	4.05	1.39872	4.35	1.47018
3.76	1.32442	4.06	1.40118	4.36	1.47247
3.77	1.32707	4.07	1.40364	4.37	1.47476
3.78	1.32972	4.08	1.40610	4.38	1.47705
3.79	1.33237	4.09	1.40854	4.39	1.47933
3.80	1.33500	4.10	1.40099	4.40	1.48160
3.81	1.33763	4.11	1 41342	4.41	1.48387
3.82	1.34025	4.12	1.41585	4.42	1.48614
3.83	1.34286	4.13	1.41828	4.43	1.48840
3.84	1.34547	4.14	1.42070	4.44	1.49065
3.85	1.34807	4.15	1.42311	4.45	1.49290
3.86	1.35067	4.16	1.42552	4.46	1.49535
3.87	1.35325	4.17	1.42792	4.47	1.49739
3.88	1.35584	4.18	1.43031	4 48	1.49962
3 89	1.35841	4 19	1.43270	4.49	1.50185
3 90	1.36098	4.20	1.43508	4.50	1.50408
3 91	1.36354	4.21	1.43746	4.51	1.50630
3.92	1.36609	4.22	1.43984	4.52	1.50851
3.93	1.36864	4.23	1.44220	4.53	1.51072
3.94	1.37118	4.24	1.44456	4.54	1.51293
3 95	1.37372	4 25	1.44692	4.55	1.51513
3.96	1.37624	4.26	1.44927	4.56	1.51732
3.97	1.37877	4.27	1.45161	4.57	1.51951
3.98	1.38128	4.28	1.45395	4.58	1.52170
3.99	1.38379	4.29	1.45629	4.59	1.52388
4.00	1.38629	4.30	1.45862	4.60	1.52606

HYPERBOLIC LOGARITHMS.—Continued.

N.	Log.	N.	Log.	N.	Log.
4.61	1.52823	4.91	1.59127	5.21	1.65058
4.62	1.53039	4.92	1.59331	5.22	1.65250
4.63	1.53256	4.93	1.59534	5.23	1.65451
4.64	1.53471	4.94	1.59737	5.24	1.65632
4.65	1.53687	4.95	1.59939	5.25	1.65822
4.66	1.53902	4.96	1.60141	5.26	1.66013
4.67	1.54116	4.97	1.60342	5.27	1.66203
4.68	1.54330	4.98	1.60543	5.28	1.66393
4.69	1.54543	4.99	1.60744	5.29	1.66582
4.70	1.54756	5.00	1.60944	5.30	1.66771
4.71	1.54969	5.01	1.61144	5.31	1.66959
4.72	1.55181	5.02	1.61343	5.32	1.67147
4.73	1.55393	5.03	1.61542	5.33	1.67335
4.74	1.55604	5.04	1.61741	5.34	1.67523
4.75	1.55814	5.05	1.61939	5.35	1.67710
4.76	1.56025	5.06	1.62137	5.36	1.67896
4.77	1.56235	5.07	1.62334	5.37	1.68083
4.78	1.56444	5.08	1.62531	5.38	1.68269
4.79	1.56653	5.09	1.62728	5.39	1.68455
4.80	1.56862	5.10	1.62924	5.40	1.68640
4.81	1.57070	5.11	1.63120	5.41	1.68825
4.82	1.57277	5.12	1.63315	5.42	1.69010
4.83	1.57485	5.13	1.63511	5.43	1.69194
4.84	1.57691	5.14	1.63705	5.44	1.69378
4.85	1.57898	5.15	1.63900	5.45	1.69562
4.86	1.58104	5.16	1.64094	5.46	1.69745
4.87	1.58309	5.17	1.64287	5.47	1.69928
4.88	1.58515	5.18	1.64481	5.48	1.70111
4.89	1.58719	5.19	1.64673	5.49	1.70293
4.90	1.58924	5.20	1.64866	5.50	1.70475

HYPERBOLIC LOGARITHMS —Continued.

N.	Log.	N.	Log.	N.	Log.
5.51	1.70656	5.81	1.75958	6.11	1.80993
5.52	1.70838	5.82	1.76130	6.12	1.81156
5.53	1.71019	5.83	1.76302	6.13	1.81319
5.54	1.71199	5.84	1.76473	6.14	1.81482
5.55	1.71380	5.85	1.76644	6.15	1.81645
5.56	1.71560	5.86	1.76815	6.16	1.81808
5.57	1.71740	5.87	1.76985	6.17	1.81970
5.58	1.71919	5.88	1.77156	6.18	1.82132
5.59	1.72098	5.89	1.77326	6.19	1.82294
5.60	1.72277	5.90	1.77495	6.20	1.82455
5.61	1.72455	5.91	1.77665	6.21	1.82616
5.62	1.72633	5.92	1.77834	6.22	1.82777
5.63	1.72811	5.93	1.78002	6.23	1.82938
5.64	1.72988	5.94	1.78171	6.24	1.83098
5.65	1.73166	5.95	1.78339	6.25	1.83258
5.66	1.73342	5.96	1.78507	6.26	1.83418
5.67	1.73519	5.97	1.78675	6.27	1.83578
5.68	1.73695	5.98	1.78842	6.28	1.83737
5.69	1.73871	5.99	1.79009	6.29	1.83896
5.70	1.74047	6.00	1.79176	6.30	1.84055
5.71	1.74222	6.01	1.79342	6.31	1.84214
5.72	1.74397	6.02	1.79509	6.32	1.84372
5.73	1.74572	6.03	1.79675	6.33	1.84530
5.74	1.74746	6.04	1.79840	6.34	1.84688
5.75	1.74912	6.05	1.80006	6.35	1.84845
5.76	1.75094	6.06	1.80171	6.36	1.85003
5.77	1.75267	6.07	1.80336	6.37	1.85160
5.78	1.75440	6.08	1.80500	6.38	1.85317
5.79	1.75613	6.09	1.80665	6.39	1.85473
5.80	1.75786	6.10	1.80829	6.40	1.85630

HYPERBOLIC LOGARITHMS.—Continued.

N.	Log.	N.	Log.	N.	Log.
6.41	1.85786	6.71	1.90360	7.01	1.94734
6.42	1.85942	6.72	1.90509	7.02	1.94876
6.43	1.86097	6.73	1.90658	7.03	1.95019
6 44	1.86253	6.74	1.90806	7 04	1.95161
6.45	1.86408	6.75	1.90954	7.05	1.95303
6.46	1.86563	6.76	1.91102	7.06	1 95445
6.47	1.86718	6.77	1.91250	7.07	1.95586
6.48	1.86872	6 78	1.91398	7.08	1.95727
6.49	1.870.6	6.79	1.91545	7.09	1.95869
6.50	1.87180	6.80	1.91692	7.10	1.96009
6.51	1.87334	6.81	1.91839	7.11	1 96150
6.52	1.87487	6.82	1.91986	7.12	1.96291
6.53	1.87641	6.83	1.92132	7.13	1 96431
6.54	1.87794	6.84	1.92279	7.14	1.96571
6.55	1.87947	6.85	1.92425	7.15	1.96711
6.56	1.88099	6.86	1.92571	7.16	1.96851
6.57	1.88251	6.87	1.92716	7.17	1.96991
6.58	1.88403	6.88	1.92862	7.18	1.97130
6.59	1.88555	6.89	1.93007	7.19	1.97269
6.60	1.88707	6.90	1.93152	7.20	1.97408
6.61	1.88858	6.91	1.93297	7.21	1.97547
6.62	1.89010	6.92	1.93442	7.22	1.97685
6.63	1.89160	6.93	1.93586	7.23	1 97824
6.64	1.89311	6.94	1.93730	7.24	1.97962
6.65	1.89462	6.95	1.93874	7.25	1.98100
6.66	1.89612	6.96	1.94018	7.26	1.98238
6.67	1.89762	6.97	1.94162	7.27	1.98376
6.68	1.89912	6.98	1.94305	7 28	1.98513
6.69	1.90061	6.99	1.94448	7.29	1.98650
6.70	1.90210	7.00	1.94591	7.30	1.98787

HYPERBOLIC LOGARITHMS.—Continued.

N.	Log.	N.	Log.	N.	Log.
7.31	1.98924	7.61	2.02946	7.91	2.06813
7.32	1.99061	7.62	2.03078	7.92	2.06939
7.33	1.99198	7.63	2.03209	7.93	2.07065
7.34	1.99334	7.64	2.03340	7.94	2.07191
7.35	1.99470	7.65	2.03471	7.95	2.07317
7.36	1.99606	7.66	2.03601	7.96	2.07443
7.37	1.99742	7.67	2.03732	7 97	2.07568
7.38	1.99878	7.68	2.03862	7.98	2.07694
7.39	2.00013	7.69	2.03993	7.99	2.07819
7.40	2.00148	7.70	2.04122	8.00	2.07944
7.41	2.00283	7.71	2.04252	8.01	2.08069
7.42	2.00418	7.72	2.04381	8.02	2.08194
7.43	2.00553	7.73	2.04511	8.03	2.08318
7.44	2.00687	7.74	2.04640	8.04	2.08443
7 45	2.00821	7.75	2.04769	8.05	2.08567
7 46	2 00956	7.76	2.04898	8.06	2 08691
7 47	2.01089	7.77	2.05027	8.07	2.08815
7.48	2.01223	7.78	2.05156	8.08	2.08939
7.49	2.01357	7.79	2.05284	8.09	2.09063
7.50	2.01490	7.80	2.05412	8.10	2.09186
7.51	2.01624	7.81	2.05540	8.11	2.09310
7.52	2.01757	7.82	2.05668	8.12	2.09433
7.53	2.01890	7.83	2.05796	8 13	2.09556
7.54	2.02022	7.84	2.05924	8.14	2.09679
7.55	2.02155	7.85	20.6051	8.15	2.09882
7.56	2.02287	7.86	2.06179	8.16	2.09924
7.57	2.02419	7.87	2.06306	8.17	2.10047
7.58	2.02551	7.88	2.06433	8.18	2.10169
7.59	2.02683	7.89	2.06560	8.19	2.10291
7.60	2.02815	7.90	2.06686	8.20	2.10413

GUIDE POSTS ON THE ENGINEER'S JOURNEY. 175

HYPERBOLIC LOGARITHMS.—Continued.

N.	Log.	N.	Log.	N.	Log.
8.21	2.10535	8.51	2.14124	8.81	2.17589
8.22	2.10657	8.52	2.14242	8.82	2.17702
8.23	2.10779	8.53	2.14359	8.83	2.17815
8.24	2.10900	8.54	2.14476	8.84	2.17929
8.25	2.11021	8.55	2.14593	8.85	2.18042
8.26	2.11142	8.56	2.14710	8.86	2.18155
8.27	2.11263	8.57	2.14827	8.87	2.18267
8.28	2.11384	8.58	2.14943	8.88	2.18380
8.29	2.11505	8.59	2.15060	8.89	2.18492
8.30	2.11626	8.60	2.15176	8.90	2.18605
8.31	2.11750	8.61	2.15292	8.91	2.18717
8.32	2.11866	8.62	2.15409	8.92	2.18830
8.33	2.12986	8.63	2.15524	8.93	2.18942
8.34	2.12006	8.64	2.15641	8.94	2.19054
8.35	2.12226	8.65	2.15756	8.95	2.19165
8.36	2.12346	8.66	2.15871	8.96	2.19277
8.37	2.12465	8.67	2.15987	8.97	2.19389
8.38	2.12585	8.68	2.16102	8.98	2.19500
8.39	2.12704	8.69	2.16217	8.99	2.19611
8.40	2.12823	8.70	2.16332	9.00	2.19722
8.41	2.12942	8.71	2.16447	9.01	2.19834
8.42	2.13061	8.72	2.16562	9.02	2.19944
8.43	2.13180	8.73	2.16677	9.03	2.20055
8.44	2.13298	8.74	2.16791	9.04	2.20166
8.45	2.13417	8.75	2.16905	9.05	2.20277
8.46	2.13535	8.76	2.17020	9.06	2.20387
8.47	2.13653	8.77	2.17134	9.07	2.20497
8.48	2.13771	8.78	2.17248	9.08	2.20607
8.49	2.13889	8.79	2.17361	9.09	2.20717
8.50	2.14007	8.80	2.17475	9.10	2.20827

HYPERBOLIC LOGARITHMS.—Continued.

N.	Log.	N.	Log.	N.	Log.
9.11	2.20937	9.41	2.24177	9.71	2.27316
9.12	2.21047	9.42	2.24284	9.72	2.27419
9.13	2.21157	9.43	2.24390	9.73	2.27521
9.14	2.21266	9.44	2.24496	9.74	2.27624
9.15	2.21375	9.45	2.24601	9.75	2.27727
9.16	2.21485	9.46	2.24707	9.76	2.27829
9.17	2.21594	9.47	2.24813	9.77	2.27932
9.18	2.21703	9.48	2.24918	9.78	2.28034
9.19	2.21812	9.49	2.25024	9.79	2.28136
9.20	2.21920	9.50	2.25129	9.80	2.28238
9.21	2.22029	9.51	2.25234	9.81	2.28340
9.22	2.22137	9.52	2.25339	9.82	2.28442
9.23	2.22246	9.53	2.25444	9.83	2.28544
9.24	2.22354	9.54	2.25549	9.84	2.28646
9.25	2.22462	9.55	2.25654	9.85	2.28747
9.26	2.22570	9.56	2.25759	9.86	2.28849
9.27	2.22786	9.57	2.25863	9.87	2.28950
9.28	2.22894	9.58	2.25968	9.88	2.29051
9.29	2.23001	9.59	2.26072	9.89	2.29152
9.30	2.23101	9.60	2.26176	9.90	2.29253
9.31	2.23109	9.61	2.26280	9.91	2.29354
9.32	2.23216	9.62	2.26384	9.92	2.29455
9.33	2.23324	9.63	2.26488	9.93	2.29556
9.34	2.23431	9.64	2.26592	9.94	2.29657
9.35	2.23538	9.65	2.26696	9.95	2.29757
9.36	2.23645	9.66	2.26799	9.96	2.29858
9.37	2.23751	9.67	2.26903	9.97	2.29958
9.38	2.23858	9.68	2.27006	9.98	2.30058
9.39	2.23965	9.69	2.27109	9.99	2.30158
9.40	2.24071	9.70	2.27213	10.00	2.30259

HYPERBOLIC LOGARITHMS.—Continued.

N.	Log.	N.	Log.	N.	Log.
10.25	2.32728	14.00	2.63906	21.00	3.04452
10.50	2.35137	14.25	2.65445	22.00	3.09104
10.75	2.37490	14.50	2.67415	23.00	3.03549
11.00	2.39789	14.75	2.69124	24.00	3.17805
11.25	2.42037	15.00	2.71035	25.00	3.21887
11.50	2.44235	15.50	2.74084	30.00	3.37817
11.75	2.46385	16.00	2.77512	40.00	3 68887
12.00	2.48491	16.50	2.80336	50.00	3.91202
12.25	2.50553	17.00	2.83321	60.00	4.09434
12.50	2.52573	17.50	2.86220	70.00	4.24849
12.75	2.54553	18.00	2.89037	80.00	4.38203
13.00	2.56264	18.50	2.91754	90.00	4.49981
13.25	2.58400	19.00	2.94444	100.00	4.60287
13.50	2.60269	19.50	2.97415	1000.0	6.90776
13.75	2.62104	20.00	2.99573	10000.	9.21034

TABLE SHOWING WATER AND COAL REQUIRED FOR STEAM-POWER.

H. P.	Water in gals. per hour.	Coal required in lbs. per hour.	Water in gals. per day of 10 hours.	Coal in lbs. per day of 10 hours.
5	20	20	200	200
10	41	40	410	400
15	58	60	580	600
20	72	80	720	800
25	90	100	900	1,000
30	110	120	1,100	1,200
40	145	160	1,450	1,600
50	180	200	1,800	2,000
60	220	240	2,200	2,400
70	260	280	2,600	2,800
80	290	320	2,900	3,200
100	405	400	4,050	4,000
125	450	500	4,500	5,000
150	590	600	5,900	6,000
200	725	800	7,250	8,000
250	900	1,000	9,000	10,000

SHRINKAGE OF CASTINGS.

Iron, small cylinders .... = $\frac{1}{8}$ inch per foot.
" pipes .... = $\frac{1}{8}$ inch per foot.
" girders, beams, etc. .... = $\frac{1}{8}$ inch in 15 inches.
" large cylinders, the contraction of diameter at top ...... = $\frac{1}{8}$ inch per foot.
" large cylinders, the contraction of diameter at bottom ...... = $\frac{1}{16}$ inch per foot.
" large cylinders, contraction in length .... = $\frac{1}{8}$ inch in 16 inches.
Brass, thin .... = $\frac{1}{8}$ inch in 9 inches.
" thick .... = $\frac{1}{8}$ inch in 10 inches.
Zinc .... = $\frac{5}{16}$ inch in a foot.
Lead .... = $\frac{5}{16}$ inch in a foot.
Copper .... = $\frac{3}{16}$ inch in a foot.
Bismuth .... = $\frac{5}{32}$ inch in a foot.

MELTING POINTS OF METALS AND SOLIDS.

	Fahr.			Fahr.
Antimony melts at	951°	Platinum melts at		4580°
Bismuth "	476°	Potassium "		135°
Brass "	1900°	Saltpeter "		600°
Cadmium "	602°	Steel "		2340° to 2520°
Cast Iron "	1890° to 2160°	Sulphur "		225°
Copper "	1890°	Silver "		1250°
Glass "	2377°	Tin "		420°
Gold "	2250°	Wrought Iron "		2700° to 2880°
Lead "	594°	Zinc "		740°
Ice "	32°	Aluminum "		1260°

BOILING POINTS.

	Fahr.		Fahr.
Ether	100°	Naphtha	186°
Fresh Water	212°	Oil of Turpentine	304°
Linseed Oil	640°	Salt Water	213.2°
Mercury	662°	Sweet Oil	412°

HEIGHT OF CHIMNEY IN FEET.

Horse Power of Boilers.	50			60			70			80			90		
	Area of Flue.	Side of Square Flue.	Diameter of Circular Flue.	Area of Flue.	Side of Square Flue.	Diameter of Circular Flue.	Area of Flue.	Side of Square Flue.	Diameter of Circular Flue.	Area of Flue.	Side of Square Flue.	Diameter of Circular Flue.	Area of Flue.	Side of Square Flue.	Diameter of Circular Flue.
	Sq.In.	Ins.	Ins.	Sq. In.	Ins.	Ins.	Sq. Ins.	Ins.	Ins.	Sq. Ins.	Ins.	Ins.	Sq. In.	Ins.	Ins.
25	225.0	15.0	16.92	216.75	14.7	16.61	208.5	14.44	16.29						
40	360.0	18.97	21.4	346.8	18.62	21.01	333.6	18.26	20.60						
50	450.0	21.21	23.93	433.5	20.82	23.49	417.0	20.42	23.04	400.5	20.01	22.55			
60	540.0	23.23	26.22	520.2	22.8	25.73	500.4	22.37	25.24	480.6	21.86	24.73	460.8	21.46	24.42
70	630.0	25.1	28.32	606.9	24.63	27.79	583.8	24.16	27.26	560.7	23.68	26.71	537.6	23.12	26.16
80	720.0	26.83	30.17	693.6	26.33	29.71	667.2	25.83	29.14	640.8	25.31	28.56	614.4	24.78	27.96
90	810.0	28.46	32.11	780.3	27.93	31.51	741.6	27.23	30.72	720.9	26.85	30.29	691.2	26.29	29.66
100	900.0	30.0	33.85	867.0	29.44	33.22	834.0	28.87	32.58	801.0	28.3	31.92	768.0	27.72	31.26
125							1042.25	32.28	36.42	1001.25	31.64	35.7	960.0	30.98	33.09
150										1201.5	34.65	39.11	1152.0	33.94	38.29
175										1401.75	37.44	42.24	1344.0	36.66	41.36
200										1602.0	40.0	45.10	1536.0	39.19	44.22
250													1920.0	43.83	49.48
300															
400															

HEIGHT OF CHIMNEY IN FEET.—Continued.

Horse Power of Boilers.	100			120			130			140			150		
	Area of Flue.	Side of Square Flue.	Diameter of Circular Flue.	Area of Flue.	Side of Square Flue.	Diameter of Circular Flue.	Area of Flue.	Side of Square Flue.	Diameter of Circular Flue.	Area of Flue.	Side of Square Flue.	Diameter of Circular Flue.	Area of Flue.	Side of Square Flue.	Diameter of Circular Flue.
	Sq. Ins.	Ins.	Ins.	Sq. Ins.	Ins.	Ins.	Sq. Ins.	Ins.	Ins.	Sq. Ins.	Ins.	Ins	Sq. Ins.	Ins.	Ins.
25	......	......	......	......	......	......	......	......	......	......	......	......	......	......	......
40	......	......	......	......	......	......	......	......	......	......	......	......	......	......	......
50	......	......	......	......	......	......	......	......	......	......	......	......	......	......	......
60	......	......	......	......	......	......	......	......	......	......	......	......	......	......	......
70	514.5	22.68	25.59	......	......	......	......	......	......	......	......	......	......	......	......
80	588.0	24.24	27.36	......	......	......	......	......	......	......	......	......	......	......	......
90	661.5	25.73	29.02	602.1	24.53	27.68	......	......	......	......	......	......	......	......	......
100	735.0	27.11	30.59	669.0	25.86	29.18	636.0	25.21	28.45	......	......	......	......	......	......
125	918.75	30.31	34.2	836.25	28.91	32.63	795.0	28.19	31.81	......	......	......	......	......	......
150	1102.5	33.2	37.46	1003.5	31.67	35.74	954.0	30.88	34.85	904.5	30.07	33.93	855.0	29.24	32.99
175	1278.75	35.76	40.35	1170.75	34.21	38.60	1113.0	33.36	37.64	1055.25	32.17	36.65	997.5	31.58	35.64
200	1470.0	38.34	43.26	1338.0	36.57	41.27	1272.0	35.66	40.24	1206.0	34.73	39.13	1140.0	33.76	38.1
250	1837.5	42.86	48.36	1672.5	40.89	46.14	1590.0	39.87	44.99	1507.5	38.69	43.81	1425.0	37.74	42.57
300	2205.0	46.95	52.98	2007.0	44.4	50.55	1908.0	43.68	49.29	1809.0	42.53	47.99	1710.0	41.34	46.66
400	2940.0	54.22	61.18	2676.0	51.73	58.36	2544.0	50.43	56.91	2412.0	49.11	57.41	2280.0	47.74	53.88

ADVERTISEMENTS.

PRINCIPAL OFFICES, EQUITABLE BUILDING,
No. 120 Broadway, New York.

OFFICERS.

WM. K. LOTHROP, President.
WILLIAM E. MIDGLEY, *Vice-President.*
RICHARD K. SHELDON, *Treasurer.*
VINCENT R. SCHENCK, *Secretary.*
THOMAS F. POWERS, *Superintendent.*

DIRECTORS.

Hon. FELIX CAMPBELL,	Engineer and Iron Merchant, New York City.
GEO. P. SHELDON,	President Phenix Insurance Company, of Brooklyn, N. Y.
Hon. WM. BRINKERHOFF,	Attorney and Counsellor, Jersey City, N. J.
JOHN H. FLAGLER,	National Tube Works Company, Pennsylvania and New York.
E. E. GEDNEY,	President North River Bank, New York.
WM. H. JACKSON,	Retired, Brooklyn, New York
ALBION K. BOLAN,	President Empire Refinery Company, New York.
E. H. APGAR,	of Apgar Company, Wholesale Grocers, New York.
D. R. SATTERLEE,	President Clinton Bank, New York City.
WM. K. LOTHROP,	President of Company.
WM. E. MIDGLEY,	Vice-President of Company.
R. K. SHELDON,	Treasurer of Company.
V. R. SCHENCK,	Secretary of Company.
T. F. POWERS,	Superintendent of Company.

THE AMERICAN IS TO-DAY THE LEADING STEAM BOILER INSURANCE COMPANY IN THE WORLD.

ITS AFFAIRS ARE CONDUCTED WITH ECONOMY.
ITS METHODS ARE BUSINESSLIKE.
ITS STABILITY IS UNQUESTIONED.
ITS INSPECTIONS ARE MOST THOROUGH,

From the date of its incorporation, in 1883, the record of this Company has been one of unremitting success. It entered the field with a cash capital of $200,000, and an unlimited fund of enterprise, for the purpose of giving real insurance at reduced rates. It had the indorsement of many prominent steam users who regarded the old plan of boiler insurance as little better than a certificate of inspection; and with this solid backing it snapped its fingers at opposition, and before the close of the year was well on its way to the front rank.

But it did not stop there. With its measures and methods sanctioned by the already large number of policy-holders who had found it to their advantage to place their risks with the AMERICAN, and with something gained by experience, the second year of the Company's work was in a much greater degree satisfactory.

From that time up to the present its advancement has been of a character that knew no failure. Each succeeding month found hundreds of steam users alive to the important fact that boiler insurance was a necessity, and that the AMERICAN was preëminently the Company to patronize; and as months lost their identity in years, the twelve-fold record, in each instance, was one to be proud of.

Technically, its place in the forefront can be credited to the fact that its form of policy covers every hazard of a boiler explosion, viz: Loss to property insured, loss to surrounding property, loss of life and injury to person—verily a blanket protection; and that its established principal of paying every loss in full without discount within ten days, has never been violated, not even in a single instance.

These features cannot fail to commend themselves to the steam user as the incarnation of that which is most vital in insurance.

THE AMERICAN'S INSPECTION.

The Inspection department of the AMERICAN is located at No. 79 John Street, N. Y. City, and has for its executive head, Thomas F. Powers, who served for twenty years as superintendent of the Municipal Bureau of Boiler Inspection in Brooklyn, and who has associated with him a corps of skilled engineers whose only duty is to carefully inspect every boiler insured by the Company four times a year, and oftener if requested by the assured, or thought necessary by the superintendent.

As soon as the assured receives our policy, arrangements are made for the first inspection, which, like the others following, is made at the convenience of the assured. In the report of the inspection the assured is advised of the condition of the boiler, and when necessary suggestions are made touching its repairs, improvements, etc.

It has been clearly demonstrated that the best means of preventing boiler explosions lies in the periodical examinations by experts who are trained to make inspections. Professor R. H. Thurston, in his recently published work on "Boiler Explosions, in Theory and Practice," plainly states that according to the best estimates which he has been able to make, *the number of explosions occurring in boilers that are uninsured and uninspected, is ten times greater than those occurring in boilers that are insured and regularly inspected.*

To insure proper inspection, the men who perform the work must be possessed of a sound knowledge, both practical and theoretical, of the construction of all classes of boilers, as well as of the various defects which occur under all conditions of working. These defects are many and varied; and it is only by careful training and experience that the ability to detect them and advise as to the best remedies can be acquired. Such training and experience being possessed by all the Inspectors of the AMERICAN STEAM BOILER INSURANCE COMPANY, it follows that examinations made by them are eminently trustworthy.

GUIDE POSTS ON THE ENGINEER'S JOURNEY.

AGENCIES AND DEPARTMENTS
OF THE
AMERICAN STEAM BOILER INSURANCE CO.

To whom all communications relative to business in their respective districts, should be addressed, and any inquiries so made will be given due and prompt attention.

RICHMOND, Va.—Thos. L. Alfriend, Agt., 1203 Main St.   M. G. Evans, Dist. Insp.
NEW ORLEANS, La.—A. A. Woods, Agt., 184 Gravier St.
PETERSBURG, Va.—E. W. Butcher, Agent.
LYNCHBURG, Va.—P. J. Otey & Co., Agents, 913 Main St.

LONG ISLAND DEPARTMENT.
JOHN R. WILMARTH, Manager.
BROOKLYN, N.Y.—John R. Wilmarth, Manager, Garfield Building.   Thomas J. Reynolds, Dist. Insp.

EASTERN NEW YORK DEPARTMENT.
W. O. LAUGHNA, Manager.
NEW YORK, N. Y.—W. O. Laughna, Manager, 120 Broadway.
ALBANY, N. Y.—S. C. Bull, Agt., 71 State St.   Wm. B. Ruggles, Jr., Dist. Insp.
SYRACUSE, N.Y.—Truair & Wyatt, Agents, 16 S. Salina St.   Thos. Higgins, Dist. Insp.
TROY, N. Y.—Knox & Mead, Agents, 253 River St.   Wm. J. Miles, Dist. Insp.

WESTERN NEW YORK DEPARTMENT.
N. H. MESSENGER, Manager.
BUFFALO, N. Y.—N. H. Messenger, M'g'r, 25 Chapin St.   C. D. Humphrey, Dist. Insp.
BATAVIA, N. Y.—S. A. Sherwin, Agent.
WREDSPORT, N. Y.—J. J. Mack, Agent.
PENN YAN, N. Y.—O. F. Randolph, Agent.
ROCHESTER, N. Y.—Ralph Butler, Agent.

PENNSYLVANIA DEPARTMENT.
J. F. HARTRANFT, Manager.
(The Pennsylvania Boiler Insurance Company of Philadelphia, of which J. F. Hartranft is President, and the American Steam Boiler Insurance Company of New York, have united their work in Pennsylvania, and are issuing joint policies, thus giving double security and being an additional safe-guard in case of disaster, to all steam users who hold joint policies in these companies.)

ALLENTOWN, Pa.—Miles L. Eckert, Agent, Hamilton St. and Centre Sq.   Joseph Burke, Dist. Insp.
EASTON, Pa.—E. H. Shawde & Co., Agents, 27 S. Third St.   Thos. Mintzer, Dist. Insp.
WILLIAMSPORT, Pa.—S. G. Van Gilder, Agent.   J. S. Duston, Dist. Insp.
PITTSBURGH, Pa.—Jas. W. Arrott, Agent, 541 Wood St.   Benj. Ford, Chief Insp., with Asst. Insp.
PHILADELPHIA, Pa.—J. F. Hartranft, Manager, 420 Library St.
PHILADELPHIA, Pa.—Etting & Co., Agents, 327 Walnut St.   Chas. S. Blake, Chief Insp., with Asst. Insp.

NEW ENGLAND DEPARTMENT.
A. B. SEELEY, General Manager.
Maine, New Hampshire, Vermont, Massachusetts and Rhode Island.

BOSTON, Mass.—A. B. Seeley, Gen'l Agt., 4 Pearl St.   Geo. H. Brown, Chief Insp. with Asst. Insps.
PROVIDENCE, R. I.—Grant & Eddy, Agents, 62 Westminster St.   Eugene Brown, Dist. Insp.
PORTLAND, Me.—Dow & Champlin.   D. K. Hayes, Dist. Insp.
BURLINGTON, Vt.—T. S. Peck, Agt., Peck's Block.   Jno. F. Molloy, Dist. Insp.
SKOWHEGAN, Me.—Griffin & Wentworth.

CONNECTICUT STATE DEPARTMENT.
R. M. DUNCAN, General Agent.
NEW HAVEN, Conn.—R. M. Duncan, Gen'l Agt., 82 Church St.   P. B. Hovey, Dist. Insp
DANBURY, Conn.—Theo. Hoyt, Agt., Old Savings Bank.
NEW BRITAIN, Conn.—Butler & Hatch, Agents.
WINSTEAD, Conn.—M. N. Griswold, Agent.
STAMFORD, Conn.—Satterlee & Swartout.

GUIDE POSTS ON THE ENGINEER'S JOURNEY. 185

NEW JERSEY DEPARTMENT.
WOOLSON & SCHENCK, Managers.
NEWARK, N. J.—Woolson & Schenck, Managers, 781 Broad St. John R. Ames, Chief Insp., with Ass't Insps.
PATERSON, N. J.—A. S. Beakes, Agent, 114 Ellison St.
HACKENSACK, N. J.—Robt. H. Wortendyke, Agent.

SOUTHWESTERN DEPARTMENT.
MARTIN COLLINS, Manager.
Missouri, Arkansas, Southern Illinois and Southern Kansas.
JOPLIN, Mo.—Calvin & Webster.
HOT SPRINGS, Ark.—Col. J. J. Sumpter.
PINE BLUFF, Ark.—W. H. Parker & Co.
TEXARKANA, Ark.—F. W. Offenhauser.
ST. LOUIS, Mo.—Martin Collins, Manager, Chamber of Commerce. Chas. H. Huff, Chief Insp., with Asst. Insps.
FORT SMITH, Ark.—Capt. Jno. Mathews.
KANSAS CITY, Mo.—Kenney, Medes & Crittenden, Agents.
METROPOLIS CITY, Ill.—W. A. McBane, Agent.

NORTHERN OHIO DEPARTMENT.
BINGHAM & DOUGLAS, General Agents.
CLEVELAND, Ohio.—Bingham & Douglas, General Agents, Superior and Bank Streets. C. B. Squires, Special Agent. H. F. Cook, Chief Inspector, with J. A. Morgan and L. T. Osborn, Insps.

SOUTHERN DEPARTMENT.
FARNHAM & STORY, General Agents.
Southern Ohio and West Virginia.
CINCINNATI, Ohio.—Farnham & Story, Gen'l Agents, 4th and Vine Sts. O. F. Wilson, Chief Insp., with Asst. Insps.
READING, Ohio.—F. H. Vorjohan, Agent.
DAYTON, Ohio.—M. F. Hoover, Agent.

WESTERN DEPARTMENT.
THACHER & VOIGHT, Managers,
Illinois, Indiana, Michigan, Wisconsin, Minnesota, Nebraska, Montana, Iowa, Kansas, Colorado, Alabama, Tennessee, and Kentucky.
CHICAGO, Ill.—Thacher & Voight, Mgrs., Phenix Building. Jno. D. Murphy, Chief Insp., with Asst. and Dist. Insps.
DETROIT, Mich.—F. H. Blackman, Agt., 115 Griswold St. J. B. Farrand, Dist. Insp.
LOUISVILLE, Ky.—Howard W. Hunter, Agt. J. E. Naylor, Dist. Insp.
MILWAUKEE, Wis.—S. W. Higgins, State Agent, address, 318 Phenix Bldg., Chicago, Ill.; Leroy M. Kelley, Dist. Insp., 340 Washington St., Milwaukee, Wis.
GRAND RAPIDS, Mich.—John F. Wilcox, State Agt., 97 Sheldon St. Stephen Christie, Dist Insp., 116 Butterworth Ave.
BIRMINGHAM, Ala.—Clark & Shepard, Agts. P. Nelson, Dist. Insp.
OMAHA, Neb.—Ringwalt Bros., Agts.; James Balrill. Dist. Insp., Barker Block.
MINNEAPOLIS, Minn.—Frank A. Scott, Special Agt., and Dist. Insp.
NASHVILLE, Tenn.—W. D. Voight, Special Agt. Hugh J. Blowie, Special Agt., and Dist. Insp.

ATLANTIC COAST DEPARTMENT.
MAURY & DONNELLY, Managers.
Maryland and Delaware.
BALTIMORE, Md.—Maury & Donnelly, Managers, Second and Halliday Sts. L. V. Faller, Special Agent, 416 Neal Building. E. F. McGinnis, Chief Insp., with Asst. Insps.

PACIFIC COAST DEPARTMENT.
ARMSTRONG, HUBBELL & PULIS, Managers.
SAN FRANCISCO, Cal.—Armstrong, Hubbell & Pulis, Managers. David Stark, Chief Insp., with Asst. Insps.

CANADIAN DEPARTMENT.
ROBERT FLAHERTY, Manager.
Dominion of Canada.
MONTREAL, Canada.—Robert Flaherty, Manager, 27 Imperial Building. C. J. Enger, Chief Insp.

GUIDE POSTS ON THE ENGINEER'S JOURNEY.

STEAM BOILER INSURANCE
ON
JOINT POLICIES IN PENNSYLVANIA.

Combined Capital, $800,000.
Combined Assets, $1,500,000.

OFFICERS PENNA. CO.	OFFICERS AMERICAN CO.
J. F. HARTRANFT, President.	WM. K. LOTHROP, President.
H. S. ECKERT, Vice-President.	WM. E. MIDGLEY, Vice-President.
LINN HARTRANFT, Secretary.	V. R. SCHENCK, Secretary.

PHILADELPHIA OFFICE, 420 Library Street.
ETTING & CO., 327 Walnut St., Philadelphia, Agents

The Pennsylvania Boiler Insurance Company of Philadelphia and the American Steam Boiler Insurance Company of New York have united their work in Pennsylvania, and are issuing joint policies under improved blanket forms, which give steam users positive protection from every loss incident to the explosion of a boiler.

The Pennsylvania Company was reorganized in 1888, and its directory comprises names amiliar to manufacturers and merchants throughout the State. The American is known everywhere as the soundest organization of its kind in existence.

The acceptance by a user of steam of the Pennsylvania-American policy carries with it the guarantee of the joint companies that by careful inspections they will greatly lessen the danger of disaster to the boilers, and that in any event they will hold themselves liable for any loss resulting from explosion.

These inspections are made periodically during the year, at intervals of three months, and ALWAYS AT THE CONVENIENCE OF THE ASSURED.

The responsibility of the joint companies is fixed and definite; the form of policy a complete blanket; the inspections thorough, and the payment of losses prompt and equitable.

Gen. John F. Hartranft, President of the Pennsylvania Company, and State Manager for the American, whose office is at No. 420 Library Street, Philadelphia, will be pleased to supply any additional information desired by those who may have the insurance of their boilers under advisement.

A. S. HATCH, President.   S. D. BREWER, General Manager.   LEVI HUSSEY, Engineer.

ECONOMY IN STEAM.

THE HUSSEY RE-HEATER AND STEAM PLANT IMPROVEMENT
COMPANY,
18 CORTLANDT STREET, NEW YORK,

PROPRIETORS OF THE HUSSEY SYSTEM OF

RE-HEATING EXHAUST STEAM,   WITHOUT
SUPER-HEATING LIVE STEAM AND ☞   COST
HEATING AIR OR WATER   FOR FUEL.

ALSO OF THE

COMPOUND AUTOMATIC FEED WATER HEATER

for Heating Feed Water to Boiling Point or to Boiler Temperature, as desired, WITHOUT COST FOR FUEL.

The Hussey System Heats EXHAUST STEAM to any desired temperature, and renders it available for all the purposes for which heat is required in Manufacturing Processes, Offices, Buildings, Apartment Houses, Hotels, etc., and circulates it without back pressure. ☞ Special attention given to Designing, Remodelling and Improving STEAM PLANTS, for obtaining the largest results in Power and Heat, with the least consumption of Fuel.

THE HUSSEY BACK PRESSURE VALVE Operates without Lever or Weights, and without any clatter or noise or hammering action.

ILLUSTRATED DESCRIPTIVE CIRCULARS SENT ON APPLICATION.

ESTABLISHED 1842.   INCORPORATED 1872.

WALWORTH MANUFACTURING CO.,
14 TO 20 OLIVER STREET, BOSTON, MASS.,

MANUFACTURERS AND DEALERS IN

STEAM USERS' MATERIALS.

WALWORTH DIE PLATES,
STILLSON AND ASHLY PIPE WRENCHS,
STANWOOD AND WALWORTH PIPE CUTTERS,

AND OTHER

TOOLS FOR STEAM AND GAS-FITTERS.

New York Office, 54 Gold Street.

J. J. WALWORTH, Pres.   G. H. GRAVES, Treas.
C. C. WALWORTH, Vice-Pres.   G. T. COPPINS, Sec'y.

THE HAZLETON BOILER CO.

MANUFACTURERS AND SOLE PROPRIETORS OF THE HAZELTON BOILER WHICH IS THAT TYPE OF WATER-TUBE BOILER FREQUENTLY CALLED THE PORCUPINE BOILER.

PATENTED IN THE UNITED STATES AND FOREIGN COUNTRIES.

At Work in all the Principal Industries.

STANDARD SIZES
25 TO 500 H. P., INCLUSIVE.

All imitations of the substantial features of this Boiler are Infringements, and will be prosecuted to the full extent of the Law.

GENERAL OFFICE:

No. 716 East Thirteenth St.,

(Works: Ave. D and Thirteenth St.)

NEW YORK, U. S. A.

The largest volume of driest steam with least consumption of fuel. Admirably adapted to the utilization of any refuse combustible material, and the waste-heat of furnaces. Absolutely safe under unusually high pressures. Light in weight. Perfect circulation. Very durable. No brick chimneys required. Accessible in every part for internal and external examination and cleaning. A quick steaming and easily managed boiler. This company is prepared to furnish plants of boilers of any desired capacity. Catalogue giving full information sent free on application. Correspondence solicited. **Please address all communications to the company.**

THE HAZELTON BOILER CO.

MANUFACTURERS AND SOLE PROPRIETORS OF THE HAZELTON BOILER, WHICH IS THAT TYPE OF WATER-TUBE BOILER FREQUENTLY CALLED THE PORCUPINE BOILER.

GENERAL OFFICE,
716 East 13th Street, New York, U. S. A.

TO MANUFACTURERS AND USERS OF STEAM BOILERS:

*A*S the so-called *Hazelton Tripod Boiler Company of Chicago is circulating misleading reports in regard to the recent decision of the United States Supreme Court in Kennedy v. Hazelton, we think proper to explain the facts, as follows:*

*I*NSTEAD *of holding any of our patents invalid, the Supreme Court simply decided that on the case presented, the Goulding patent, under which the Hazelton Tripod Boiler Company claims to operate, was void—so void in fact that a Court of Equity could not recognize it as property at all. It was only on this ground that the Court declined to compel Hazelton to assign it over to this Company.*

*W*E *quote as follows, from a copy of the decision on file at our office:*
"*A COURT OF CHANCERY CANNOT DECREE SPECIFIC PERFORMANCE OF AN AGREEMENT TO CONVEY PROPERTY WHICH HAS NO EXISTENCE, OR TO WHICH THE DEFENDANT HAS NO TITLE.*"

*T*O *say that the decision sustains any patent or claim of the Hazelton Tripod Boiler Company, is simply contrary to the facts in the case. The Court did nothing of the kind. The patent suits which we have brought against certain infringers are still pending, and will probably be decided some time during the coming spring or summer.*

THE HAZELTON BOILER COMPANY

KEEP YOUR BOILERS FREE FROM SCALE!
TRI-SODIUM-PHOSPHATE

WATER ∴ PURIFIER.

Absolutely the Only Perfect Water Purifier in Existence.

We earnestly recommend the TRI-SODIUM-PHOSPHATE WATER PURIFIER to every maker of steam as the ONLY INFALLIBLE SCALE PREVENTER ever offered. One trial will satisfy any engineer of the great value of this compound. It is the only remedy for Boiler Incrustations that acts on sound chemical principles, by precipitating the lime and magnesian salts, and all other scale forming bodies, as phosphates, which float through the water in light downy flakes, easily removed by blowing off. Pitting and corrosion are entirely prevented by its use. It keeps iron clean and bright, and is PERFECTLY HARMLESS. It neutralizes all acidity even in the most impure mine-waters, and by totally preventing the formation of scale t insures a LARGE SAVING IN FUEL, as well as increased steam capacity.

WE INVITE THE MOST SEARCHING EXAMINATION, AND COURT THE STRICTEST SCRUTINY AS TO THE VALUE OF THE TRI-SODIUM-PHOSPHATE WATER PURIFIER. Leading mechanical and chemical experts pronounce it to be UNEQUALED in the essential qualities of EFFICIENCY, ECONOMY and PERFECT HARMLESSNESS.

Send for pamphlet containing many strong testimonials, with directions for its use.
Correspondence cheerfully answered.

KEYSTONE CHEMICAL CO.,
No. 3 South Front Street, Philadelphia.

For Sale by
- WILSON & ROAKE, 261 Front Street, New York.
- GEO. F. PATTERSON & CO., 217 E. Fayette St. Baltimore.

☞ SEE OPPOSITE PAGE. ☜

The opinion of an Eminent Chemist on Tri-Sodium Phosphate as a Boiler Scale Preventive.

CHEMICAL LABORATORY OF DR. F. A. GENTH,
111 South Tenth Street.
PHILADELPHIA, May 11, 1880.

KEYSTONE CHEMICAL COMPANY.
3 South Front Street, Philadelphia, Pa.

GENTLEMEN:
In compliance with your request, I have made an examination of the sample of Tri-Sodium Phosphate, and after making numerous experiments with it, I have arrived at the conclusion that *it furnishes the* BEST *and* MOST EFFECTIVE means of separating such constituents of the water, as cause the formation of scales and incrustations in boilers. These constituents are converted into light flocculent precipitates which do not deposit as scales or incrustations, but remain suspended in the water and can easily be removed by blowing off.

Water containing lime, magnesia, etc., as bi-carbonates or sulphates, and silica, are at once deprived of these injurious constituents when boiled with the proper quantity of Tri-Sodium Phosphate.

Another important property of this salt is that it does not only act upon the injurious substances dissolved in the water, but even upon scales and incrustations already existing in the boilers, which are gradually removed by it; because it acts upon the solid carbonates of lime and magnesia, as well as upon the sulphate of lime, even in its anhydrous form (one of the worst kinds of boiler incrustations) by converting them into pulverulent and flocculent phosphates.

As it has no injurious effect upon the boilers, the use of this salt for the purposes indicated is of great importance. I remain, yours very truly,
(Signed) F. A. GENTH;

Geo. G. Lobdell, Pres. LOBDELL CAR WHEEL COMPANY,
Wm. W. Lobdell, V. Pres.
Geo. G. Lobdell, Jr., Secy. and Treas. WILMINGTON, DEL., Feb. 11, 1889.
KEYSTONE CHEMICAL COMPANY,
No. 3 South Front Street, Philadelphia, Pa.

GENTLEMEN:
At our Walton Furnace, in Wythe County, Virginia, the water is strongly impregnated with lime. We have used your Tri-Sodium Phosphate with perfect success for more than a year at that place. Boilers that would otherwise have had a scale one-quarter of an inch or more in thickness in a few months' use of this water, now have no scale after a run of nine months. We know of nothing else so good for the purpose, and shall continue to use it.

Yours respectfully,

GEO. G. LOBDELL, Pres.

KLUENTER & YEAGER.
Manufacturers of Parlor Frames.

ALLENTOWN, PA., June 12, 1889.
KEYSTONE CHEMICAL COMPANY.
Philadelphia, Pa.

GENTLEMEN:
We have been heretofore troubled very much by the formation of scale in large quantity in our boilers, and have tried unsuccessfully, various compounds to prevent and remove it.

At last we were induced to use your Tri-Sodium Phosphate in our boilers, which had quite a deposit on them at the time. We beg to say it certainly does all you claim for it: for not only did it prevent new formation, but loosened up the old scale, so that it was easily removed by hand, whereas we had to chop it out before, entailing much time and labor.

We cheerfully endorse and recommend it to all steam users.

Yours, very truly,

KLUENTER & YEAGER.

JENKINS BROS.' VALVES.

Have Keyed Stuffing Box and Disc. Removing Lock-Nut manufactured of best Steam Metal.

Interchangeable. Insist · on · having Valves stamped with our "Trade Mark" like on cut.

71 John St., New York.

54 Dearborn St., Chicago.

21 North Fifth St., Phila.

105 Milk Street, Boston.

REDUCED SIZE JENKINS BROS., ONE INCH GLOBE VALVE.

GUIDE POSTS ON THE ENGINEER'S JOURNEY. 193

HOOPES & TOWNSEND,

MANUFACTURE

MACHINE

AND CAR BOLTS,

"KEYSTONE"
BOILER
RIVETS,

SPLIT AND SINGLE KEYS,

RAILROAD TRACK BOLTS,

WASHERS, PATCH BOLTS,

COLD-
PUNCHED
SQUARE
AND
HEXAGON
NUTS.

LAG OR
WOOD
SCREWS,
TANK and
COOPERS'
RIVETS.

TRADE MARK.

PHILADELPHIA, PA.

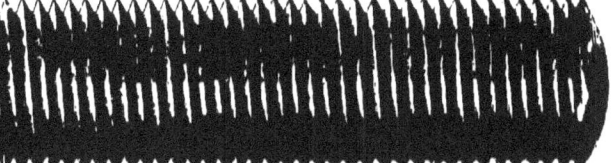

COOKE & CO.,
No. 22 Cortlandt Street, - - New York
ENGINES AND BOILERS,

HORIZONTAL

AND

UPRIGHT.

STATIONARY

AND

PORTABLE.

BECK AUTOMATIC ENGINE.

INSTALLATION OF COMPLETE STEAM PLANTS A SPECIALTY

Write us what you want and we will make low prices.

MENTION "GUIDE POST."

WEITMYER PATENT FURNACE
IGNITES ALL GASES BY INTRODUCING HOT AIR,

Secures Perfect Combustion, Consumes Smoke, Enables the use of Waste Fuels, which cannot be used in the Ordinary Furnace, Screenings, etc.

MANUFACTURED AND SOLD BY

Foundry and
Machine Department,
HARRISBURG, PA.
(Harrisburg Car M'f'g Co.)

Ide Automatic Engines,   Portable and Traction Engines,
                         Steam Road Rollers,   Boilers of all descriptions.
New York Office, - - - - - - - 17 Dey St.
Boston   "   - - - - - - - 70 Kilby St.
Baltimore "   - - - - - - - 21 So. Charles St.

THE SLOANE STEAM GENERATOR
AND
Boiler Auxiliary.

Patented October 18th, 1881; May 2d, 1882; May 16th, 1882; Re-issued December 18th, 1883

Can be attached to Old or New Boilers in TWO DAYS WITHOUT disturbing walls or changing setting.

SAVES 30 TO 50 PER CENT OF FUEL.

25 to 35 Horse Power Added to each Boiler.

200 to 250 Square Feet of Water Surface Exposed to the Same Heat as the Shell of the Boiler.

Two Boilers with this Generator Attached do the work of Three Boilers.

Eight years in use in a large number of factories. Entire satisfaction in every instance.

NO REPAIRS REQUIRED.

For price list and circulars, address

G. W. SLOANE, 53 Greenpoint Ave., Brooklyn, N. Y.

THE
BERRYMAN
PATENT
FEED WATER
HEATER
AND
PURIFIER.

● ● ●

If you are using a steam boiler and have exhaust steam there is no safer or better-paying investment than a Berryman Feed Water Heater and Purifier. It heats the feed water to the highest possible temperature with the exhaust without back pressure. It increases the steaming capacity of boilers; keeps boilers clean and free from scale; saves fuel, labor and boiler repairs. This is what we point hardest at: Our prices are no higher than other heaters of equal size and carrying capacity in gallons. To investigate will cost you nothing and may put money in you purse.

BENJ. F. KELLEY,
91 Liberty St., N. Y.

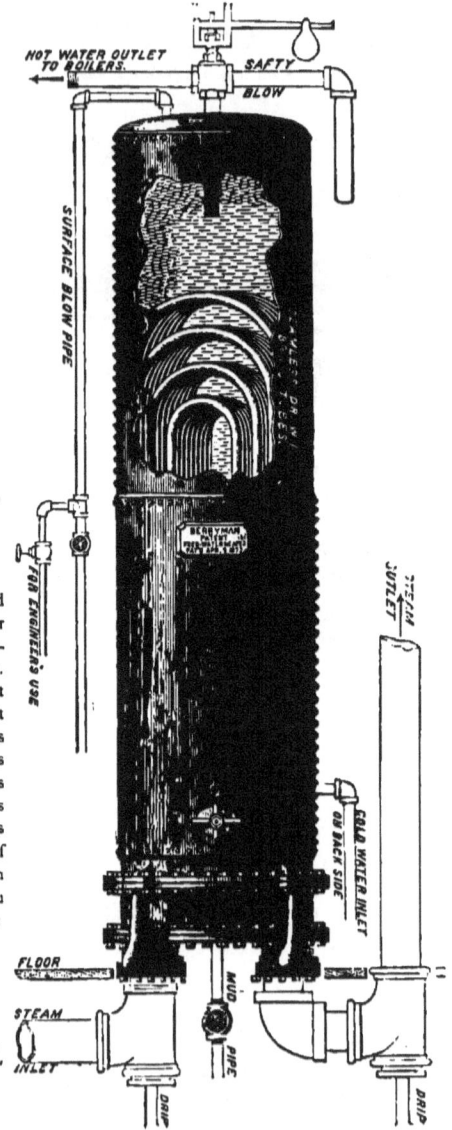

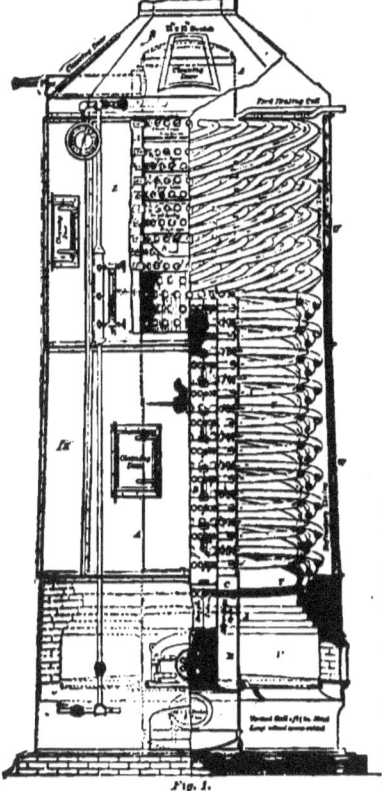

Fig. 1.
"THE 'CLIMAX' STEAM GENERATOR."

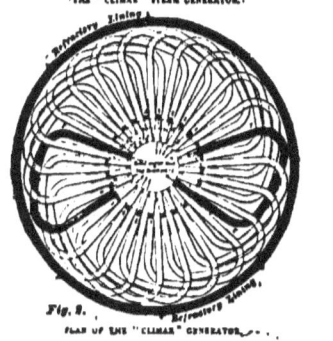

Fig. 2.
PLAN OF THE "CLIMAX" GENERATOR.

EXPLANATIONS.—*A*, Body of Boiler. *B*, Water Reservoir. *C C*, Supply Tubes. *T T*, Heating Tubes. *S*, Deflector. *V*, Fire-box. *W W*, Casing. *Z*, Refractory Lining.

THE CLIMAX
Steam Generator

Simple in Construction.

Effective in Operation.

Leakage from Expansion and Contraction Impossible.

THE ACME OF ECONOMY.—
FIFTY FEET OF HEATING
SURFACE TO EACH
FOOT OF GRATE
SURFACE.

Perfectly safe under any pressure and occupying the minimum amount of floor space. Generators of this pattern are unexcelled for stationary or marine service. Several have been in use for a number of years for both purposes, and have given thorough satisfaction. Address for further particulars,

T. F. MORRIN,
JERSEY CITY, N. J.

DANGER PREVENTED!

Your Boiler is the life of your establishment, therefore protect it!

We will remove and prevent Scale in any Steam Boiler, and ask no money until you find our remedy to work to your entire satisfaction.

WE USE NO ACID!

The Chemicals we use will neutralize Acid, Sulphur, and Mineral Waters, and will not only protect you from Boiler Scale and Oxidation of Iron, but from corrosion in every form, such as GROOVING, PITTING, and WASTING OF IRON, which causes so many explosions. If you will protect your Boilers from the above-named destructive agents, then with a reliable engineer there will be no occasion for a boiler explosion, and the money saved by the use of our Compound will more than pay the cost of it in every case.

This article has worked its way into favor entirely upon its merits, without sending out a single man to solicit orders, and we can refer to our customers in every steam-using locality from Maine to the Pacific Coast. More than thirty thousand manufacturing establishments, in the United States and Canada alone, are now using our Compound, many of whom, using a number of boilers order it by the ton.

Authors of the best works on Steam Engineering now recommend Lord's Compound as the only reliable article to prevent corrosion in boilers, and as a safe remedy for incrustation; and as a proof of this fact, we will send any of these books by mail, free of postage, at publishers' price, or free of cost with a barrel of Compound.

A VERY USEFUL BOOK ON CARE OF BOILERS FREE.

Also our circular, with full particulars of the Compound, on request.

Address, GEO. W. LORD,
316 UNION STREET, PHILAD'A, PA.

AGENTS WANTED EVERYWHERE.

HEINE SAFETY BOILER CO.

250 HORSE POWER HEINE SAFETY BOILER.

SAFE, DURABLE, ECONOMICAL.
Over 100,000 H. P. in use.

ADDRESS FOR PARTICULARS:

HEINE SAFETY BOILER CO., Bank of Commerce Bldg., ST. LOUIS.

CHICAGO.—82 Madison Street. BOSTON.—55 Oliver Street.
NEW ORLEANS.—34 St. Charles St. PHILADELPHIA.—140 S. 4th St.
DENVER.—4 Duff Block. KANSAS CITY.—1221 Union Ave.

200   GUIDE POSTS ON THE ENGINEER'S JOURNEY.

◄ E. ● J. ● WOOD, ►

CONSULTING ENGINEER AND CONTRACTOR,

243 BROADWAY   -   NEW YORK.

Superintendent of the Construction and Erection of Factories, Steam Plants, and all kinds of Machinery. Engines Indicated. Valves Adjusted and Power Measured.

COMPLETE STEAM PLANTS, BOILERS,
ENGINES OF EVERY DESCRIPTION.

SHAFTING,
   PULLEYS,
      HANGERS.

DRAWING AND SUPERINTENDENCE.   ECONOMY GUARANTEED.
          CORRESPONDENCE SOLICITED.

GOLDSMITH Patent Quick Opening and Closing Straight and Angle

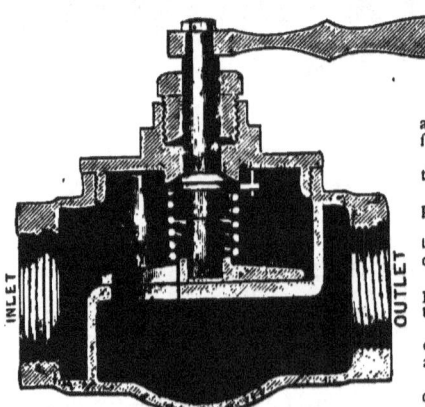

THROTTLE VALVES
And Main Stop Valves.

One movement of the lever opens, closes and graduates the steam from the finest flow to the full pipe area

The valves and seats are so constructed that uneven wear is prevented.

All parts can be taken from body and replaced without breaking pipe connections.

In their operation the valves never leave their seats; they follow and take up their own wear.

Sediments or scale from pipes cannot possibly get between the valves and seats, to cut or wear them.

The valves work positively, easily and quickly under ordinary or heavy pressure, and are always tight.

Every valve gives full discharge of pipe opening.

The valve can be instantly closed when required.

SEND FOR CIRCULAR.

WM. T. ANDREWS, 55 Oliver Street, Boston, Mass.

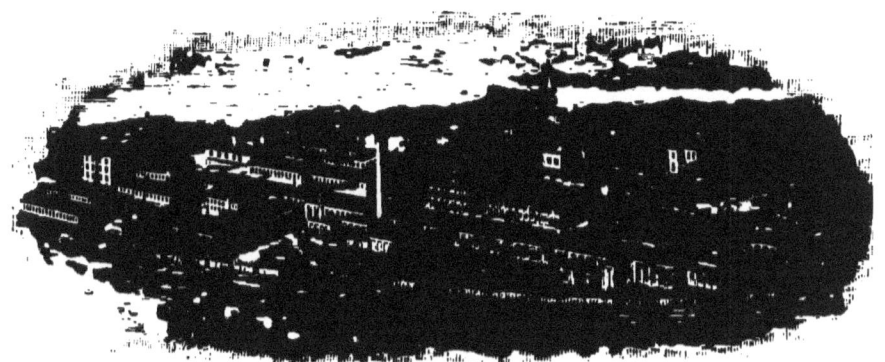

The Corliss Steam Engine Co
PROVIDENCE, R.I.
Incorporated June, 1856; Established by
GEORGE H. CORLISS,
INVENTOR OF THE CELEBRATED
 **"CORLISS ENGINE."**
DESIGNER & BUILDER OF THE FAMOUS
"CENTENNIAL ENGINE,"
Exhibited at the Philadelphia Exposition, 1876.

These works have been fully equipped, at great cost, with heavy special tools, of his invention, for the manufacture of this perfected engine, which is a guarantee of superiority in workmanship, and interchangeability of parts, never before attempted in the line of steam machinery.

The public will understand that we have no relations with American or European builders of so-called "Improved Corliss Engines," and that the final and perfected Engine of Mr. George H. Corliss, embodying his latest ideas, is to be obtained exclusively at our works.

ALSO, MANUFACTURERS OF THE

CORLISS PATENT VERTICAL TUBULAR WATER LEG BOILER,
Especially adapted for compound and triple expansion engines requiring superheated steam and at very high pressure.

DOWNIE EUCALYPTUS Boiler Scale Preventative and Remover.

PATENTED IN ALL THE PRINCIPAL COUNTRIES IN THE WORLD.

Manufactured by the DOWNIE B. I. P. CO.,

204 Market Street, San Francisco.   Branch Office, 4 Red Cross Street, Liverpool.

This extract from the leaves of the Eucalyptus Tree, has been adopted and is now in use by most of the American S. S. lines, the U. S. Navy, and the largest stationary plants in the country. Send for pamphlet containing letters from prominent engineers, directions for use, price, etc.

DELAFIELD, MORGAN, KISSELL & CO., Agents, 91 Hudson Street, New York.

THE HUB SHAKING GRATE

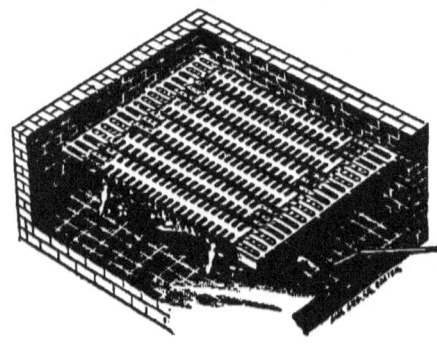

IS THE BEST AND THE CHEAPEST.

Saves more time, labor, and fuel than any other.

Don't fail to send for circular with testimonials from prominent parties to

PORTSMOUTH WRENCH CO.,

151 Congress Street,

BOSTON.

THE ÆTNA SHAKING GRATE

Superior to all others. Sixty per cent. air space. No change in F ~e necessary. Send for Circulars. Aetna Grate . :o., 110 Liberty St., New York.

THE SALAMANDER
Patent Interlocking Grate Bars

"OLD RELIABLE."
Used and approved in upwards of 10,000 Furnaces.
MADE ONLY BY THE
SALAMANDER GRATE BAR COMPANY,
110 Liberty St., New York.

JEWELL PURE WATER COMPANY,

CHICAGO, ILL.

CLAIM.	GUARANTEE.
1st. Absolute perfection in purifying and softening waters.	1st. To remove mineral and organic matter whether animal or vegetable.
2d. Simplicity of System.	2d. To soften hard water taking out both *temporary* and *permanent* hardness.
3d. Quickness in cleaning, five to ten minutes being ample time for the largest size.	3d. To remove suspended matter and to precipitate and remove matter held in solution.
4th. Positive assurance that the filter bed is perfectly clean after each washing.	4th. To remove alkali, rendering such waters soft, pleasant and agreeable.
5th. Durability.—There being no springs, complicated valves or delicate parts to soon wear off.	5th. Water that has been filtered by our system for steam-boiler use will absolutely prevent all incrustations forming, and prevent foaming or priming.
6th. Strength in Construction—being practically indestructible.	6th. Water that we treat will be bright, clean soft, sparkling and wholesome.
7th. Steady and regular feeding of coagulating or precipitating chemical.	**NO PAYMENT REQUIRED** until guarantee is fulfilled to satisfaction of purchaser.
8th. The *only sight-feed* chemical apparatus.	

MORISON, ALLEN & CO., Eastern Agents, 145 Broadway, New York.

THE ARMSTRONG MANUFACTURING CO.,
BRIDGEPORT, CONN.
MANUFACTURERS OF THE ARMSTRONG CELEBRATED

STOCKS  AND DIES,

AND OTHER TOOLS FOR WATER, GAS, AND STEAM FITTERS' USE.
We especially recommend our NEW PLAIN and HINGED VISES, PIPE CUTTERS and WRENCHES
SEND FOR CATALOGUE AND PRICES.

STEAM-PIPE AND BOILER COVERING.
MAGNESIA.
CAN BE PUT ON BY ANYBODY.

FIRE  PROOF.

ROBERT A. KEASBEY & CO., 58 Warren Street, New York.

R. F. HAWKINS, Proprietor.   C. H. MULLIGAN, Supt.   E. B. JENNINGS, Engineer.

R. F. HAWKINS IRON WORKS.

IRON AND WOODEN BRIDGES AND TURN
TABLES, STEAM BOILERS, IRON CASTINGS, BOLTS, Etc.

SPRINGFIELD, MASS.

NEWBURGH STEAM BOILER WORKS.

P. DELANY.

ESTABLISHED 1870.

HORIZONTAL TUBULAR BOILERS.

There being more of this style of boiler used than any other, we have endeavored to present a list that will suit any purchaser's desire, although we have made many sizes not found on the list.

All boilers, from 20 horse-power upward, are double riveted in the horizontal seams and dome where connected to shell. The tubes are spaced one inch apart both horizontally and vertically, except middle vertical row, where the space is two inches. No tube is nearer the side of shell than three inches, and nine inches from the bottom in the larger sizes.

The larger steel mills are making very large plates or us, and we are enabled to have our plate extend around the boiler and place the seam above the top of the lugs and out of the fire. This is a great advantage as not having two thicknesses of double riveting of steel together in the fire, and also, in case of repairing, it is more accessible.

The fixtures furnished with these boilers are—heavy cast iron front with double flue, furnace and ashpit doors and linings; arch plates; front, middle (where grates are in two lengtsh), and back grate bar bearers; grate bars; ash door and frame, with bolts; lug plates and friction rollers; brick stays; anchor and brick stay bolts and nuts; safety valve; steam gauge; glass water gauge; gauge cocks; water column; blow-off, feed and check valves.

These boilers are made of best materials throughout, strongly braced with crow-foot braces, for 100 lbs. steam pressure.

A suitable hand hole, with plate and guard, is put in each tube-head.

Each boiler is carefully tested to 150 pounds warm water pressure, and made tight before shipment.

A line of these boilers, from 25 to 100 horse power, constantly on hand.

ROBERT POOLE & SON CO.,

LEFFEL TURBINE
WATER WHEEL,

Made of Best Materials, and in the Best Style of Workmanship.

Machine Molded Mill Gearing,

From one to twenty feet diameter, of any desired face or pitch, molded by our own Special machinery.

SHAFTING, PULLEYS, AND HANGERS

Of the Latest and Most Improved Designs.

STEAM ENGINES, BOILERS,

MIXERS AND GENERAL OUTFIT FOR FERTILIZER WORKS.

Shipping facilities the best in all directions.   N. B.—Special Attention given to Heavy Gearing.

ROBERT POOLE & SON CO., Baltimore, Md.

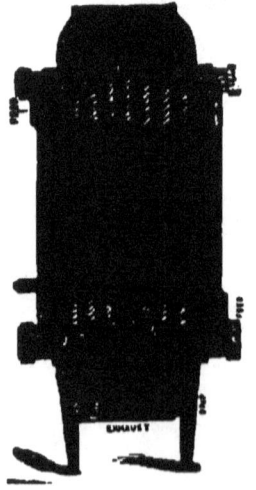

THE WAINWRIGHT
CORRUGATED COPPER TUBE HEATER.

OVER 60,000 H. P. IN USE.

Highest results obtained with exhaust steam alone, without back pressure.
The most carefully constructed Heater in the market.
Coil, Vertical, and Horizontal Straight Tube Heaters.
Heaters for Compound Condensing Engines, Expansion Joints, Filters, and Condensers.   Prices very low.
Heaters constructed of Cast Iron and Steel Shells, Copper Tubes, Brass Connections.   No Wrought Iron to rust out.

SEND FOR NEW ILLUSTRATED CATALOGUE.

THE WAINWRIGHT MFG. CO.
OF MASS.,

Factory, Medford.   **34 OLIVER ST., BOSTON.**

THE ALBANY STEAM TRAP CO.
Office and Works, 78 and 80 CHURCH STREET, ALBANY, N. Y.
SEND FOR CIRCULARS.

RETURN STEAM TRAPS AND PUMP GOVERNORS
Automatically drain the water of condensation from HEATING COILS and return it to the boilers, whether the coils are above or below the water level in boiler.
SPECIAL DUPLEX STEAM PUMPS for Boiler feeding and Pumping Condensation direct from Heating Coils, at any temperature, also for other duties.

THE ALBANY STEAM TRAP CO.
Office and Works, 78 and 80 CHURCH STREET, ALBANY, N. Y.
BOILER PURIFIER.

DUPLEX WATER FILTERS.
For filtering and purifying water for City Supply, Manufacturing Purposes, Breweries, Hotels, Laundries, Household Use, Etc.

RENEWABLE SEAT GATE STOP AND CHECK VALVES.
The body of Valve never removed from Pipe to renew Seats or Gates.

This Apparatus Removes the Impurities from Steam Boilers, by the process of a *continous circulation* of the water from the *Boiler*, through the *Filter*, and back into the *Boiler*. The scale forming impurities that are held in suspension are thus brought in contract with "and arrested" by the filtering agent contained in the *Filter*, while under *pressure*, and at a *temperature* limited only by that contained in the *Boiler*. By this method we have the advantage of *precipitation* not obtainable, by first passing the feed water through a heater and filter, before entering the boiler.

THE SCHAFFER & BUDENBERG
IMPROVED EXHAUST STEAM INJECTOR.

Adapted for working at Pressures up to 120 lbs. and higher.

These Injectors are working with great success on Stationary Engines and Boilers, as also on Steamers, Tugs, Dredges, etc., as they work WELL in the roughest weather.

The ECONOMY In FUEL shown over the ORDINARY LIVE STEAM INJECTOR, has never been found LESS THAN 20 PER CENT in actual practice.

Schaffer & Budenberg,
40 John St., New York.

Sole licensees and manufacturers in the United States.

Branch Office,
18 South Canal St., Chicago.

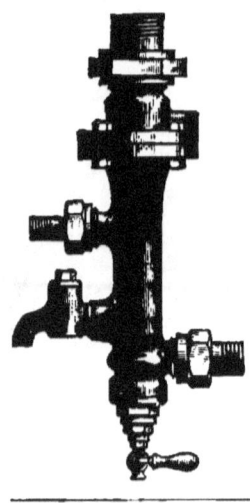
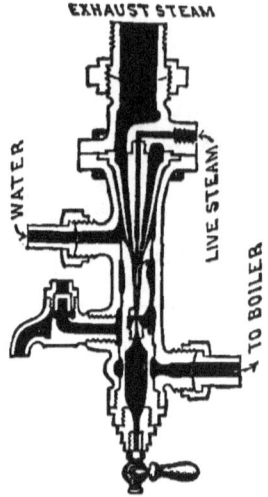

THE HOTCHKISS AUTOMATIC BOILER CLEANER

When applied to a STEAM BOILER will remove all MUD, SEDIMENT, LIME, or OIL, and prevent FOAMING, SCALING and BURNING. Will do more for $75.00 than any device extant can possibly do at any price. By keeping the water purified saves FUEL, LABOR, and REPAIRS.

Beware of Infringers.

NO CLEANING DAYS.
COMPOUNDS.
DETENTIONS.
REPAIRS.
COST, EXCEPT FIRST.
SIMPLE AND EFFECTIVE.

"**HOW TO KEEP BOILERS CLEAN**" Is the title of a 96 page illustrated book giving a full description of the cleaner, with list of users and testimonials, mailed free to any address, by

JAS. H. HOTCHKISS,
120 LIBERTY STREET, NEW YORK.

N. B.—A special size for very large boilers, or boilers using Artesian or well water highly charged with sulphate of lime.

GUIDE POSTS ON THE ENGINEER'S JOURNEY. 209

THE STANDARD

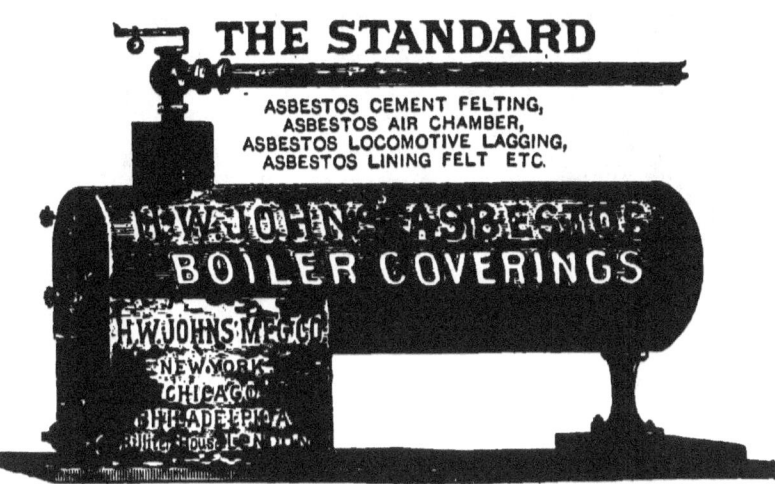

ASBESTOS CEMENT FELTING,
ASBESTOS AIR CHAMBER,
ASBESTOS LOCOMOTIVE LAGGING,
ASBESTOS LINING FELT ETC.

W E are the inventors and sole manufacturers of the following articles, and are the most extensive manufacturers in the world of materials for the purposes designated, all of which are ready for use:

H. W. JOHNS' GENUINE ASBESTOS MATERIALS.

ASBESTOS PISTON ROD PACKING.
ASBESTOS WICK PACKING.
ASBESTOS MILL-BOARD OR FLAT PACKING.
ASBESTOS GASKETS AND RINGS.
ASBESTOS AND RUBBER CLOTH, TAPE, ETC.
ASBESTOS CLOTH, ROPE, TWINE, ETC.

ASBESTOS ROOFING.
ASBESTOS ROOF COATING.
ASBESTOS ROOF PAINTS.
ASBESTOS CEMENT.
ASBESTOS BUILDING FELT.
ASBESTOS SHEATHING.

ASBESTO-SPONGE SECTIONAL PIPE COVERING.

H. W. Johns' ASBESTOS Liquid Paints

STRICTLY PURE COLORS IN OIL, ETC.

Descriptive Price Lists and Samples Free, by mail.

H. W. JOHNS M'F'G CO,

87 MAIDEN LANE, NEW YORK.

240 & 242 RANDOLPH ST : CHICAGO.   170 & 172 NORTH 4TH ST., PHILADELPHIA.
44 OLIVER STREET, BOSTON.

Send for Catalogues and Price Lists.

THE AMERICAN WATCHMAN'S
TIME · DETECTOR.

SEND FOR CIRCULARS AND LIST OF PATRONS.

PUT UP UPON GUARANTEE.

ONLY CONCERN EXCLUSIVELY IN THIS BUSINESS.

J. S. MORSE, Superintendent,

234 BROADWAY, - NEW YORK.

IMPERIAL CHEMICAL CO.
FOR THE MANUFACTURE OF
BOILER COMPOUNDS,
Office and Laboratory, 324 St. John Street,

A. WASSERMAN, Manager.   PHILADELPHIA, PA.

Boilers Cleaned and Kept Clean, by Contract per Year.
SATISFACTION GUARANTEED.

Parties sending specimens of their Scale (about one oz.) will enable us, by analyzing it, to make a Compound specially adapted to their requirements.

IMPORTANT TO STEAM USERS.

It is now more than twenty years since "Lord's Boiler Cleansing Compound" was first placed upon the market, and it is now in regular use in more than 30,000 establishments in the United States and elsewhere. The sole manufacturer of this compound is now, and always has been, Mr. GEORGE W. LORD. of 316 Union street, Philadelphia. There is no question but the genuine "Lord's Boiler Compound " is by far the best article in the world of its class. It is indorsed by all the leading authorities upon such matters, including the leading works upon engineering and the mechanical journals of the country. Edwards, in his "American Steam Engineer," (1889), says: "Lord's Compound is probably the only one of its kind that is unanimously indorsed by professional men throughout the length and breadth of this continent, among whom are authors of mechanical books, engineers in charge of works, practical chemists, professional inspectors, and manufacturers having large capital invested in steam boilers. This unanimous indorsement is probably due to the fact that not a single accident has occurred to any boiler having the Compound in use, though we find the article in use in every steam-using locality from Canada to Mexico and from Maine to the Pacific slope." And Stephen Roper, M. E., says in his "Engineer's Handy Book," that " Lord's Boiler Compound appears to be the only chemical preparation in use at present that will prevent the formation of scale or remove it after it has been formed, in any class of boilers, whether stationary, locomotive or marine, as it neutralizes the action of the natural chemical salts which form the basis of all scale and incrustation."

The choice of any of the above valuable books on Steam Engineering and its branches will be mailed free of charge to any engineer ordering a barrel of compound direct from G. W. LORD, Sole Manufacturer of Lord's Compound, No. 316 Union Street, Philadelphia.

PHENIX INSURANCE COMPANY,

BROOKLYN, N.Y.

CASH CAPITAL, $1,000,000.

GROSS ASSETS, JUNE, 29, 1889, - $4,573,249.93.

LIABILITIES, - - - - - - 3,343,944.40.

SURPLUS AS TO POLICY HOLDERS, $1,229,305.53.

LOSSES PAID SINCE ORGANIZATION, $36,764,930.98.

212   GUIDE POSTS ON THE ENGINEER'S JOURNEY.

THE WHITEHILL CORLISS ENGINE.
Patented April 10, 1888.

BUILT BY THE NEWBURGH STEAM ENGINE WORKS.
ROBERT WHITEHILL. Works, Newburgh, N. Y., U. S. A. SEND FOR DESCRIPTIVE PAMPHLET.

NEW YORK OFFICE, No. 6 Coal and Iron Exchange, corner Cortlandt and Church Streets.

WHITEHILL CORLISS ENGINE.
PATENTED APRIL 10, 1888.

ROBERT WHITEHILL, - - Newburgh, N. Y.
NEWBURGH STEAM ENGINE WORKS.

E. A. WILDT & CO.,

MANUFACTURERS OF

ELECTRICAL APPLIANCES,

ESTIMATES FURNISHED.

SPECIAL APPARATUS DESIGNED.

83 MURRAY STREET, NEW YORK.

ENGINEERS AND STEAM USERS

Who want Clean Boilers, and no risk of Explosion, would do well to write to

S. W. LORD & CO.,

11 SOUTH 9TH STREET,   PHILADELPHIA, PA.,

for circulars of price and other particulars of their

BOILER COMPOUND,
(Lord's Patent.)

Special notice to Engineers who are desirous of passing examinations.—A member of this firm (Practical Engineer) will give advice free.

YOU KNOW

SWEET, ORR & CO.,

make Overalls, one pair of which will out-wear two pair of ordinary make. Read these facts about the Pants they make: $1.00 buys a good, durable pair of Working Pants. $1.25, $1.50, $1.75 buys Pants of finer quality and more stylish appearance. $2.00, $2.50. $3.00, $3.50 buys a handsome pair of wool mixed, or all wool Pants that you would be proud of.

PANTS made by SWEET, ORR & CO., fit perfectly, and are the correct shape. The goods are selected for their durability, and no trashy materials are used.

There is a great deal of satisfaction in wearing a pair of Pants until completely worn out, without having a rip or buttons come off, or button-holes tear.

SWEET, ORR & CO.'S PANTS will give you that satisfaction.

Try a pair and you will never buy any other make.

Be sure to see that Sweet, Orr & Co. is on the buttons; none others are genuine. Ask your dealer for them, and don't let him palm off an inferior make.

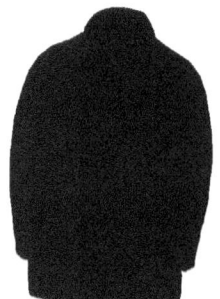

Don't fail to look at SWEET, ORR & CO.'S

LINED DUCK COATS,

if you want to keep warm and dry in the winter.

SWEET, ORR & CO.,
NEWBURGH, N. Y.

115 WORTH STREET, NEW YORK CITY.

229 TO 235 FRANKLIN STREET, CHICAGO, ILL.

BUY THE BEST !

MORE THAN 1000 IN USE.

The Smith Feed-Water Heater & Purifier Co.,

GUARANTEE FOR THEIR DEVICE:

The positive and effective removal of all impurities contained in water.

The absolute avoidance of cold water coming in contact with hot surface of the boiler.

To supply feed-water to boilers at a temperature of more than 300 degrees Fahrenheit.

A CLEAN BOILER OR NO PAY.

Send for pamphlets giving full particulars, price, etc.

EASTERN DEPARTMENT,

SMITH FEED-WATER HEATER & PURIFIER CO.,

81 TIMES BUILDING, NEW YORK.

GENERAL OFFICE,

SMITH FEED-WATER HEATER & PURIF...

No. 1 N. BROADWAY   ...UIS.

S. T. ...LEYER, General ...ager.

www.ingramcontent.com/pod-product-compliance
Lightning Source LLC
Chambersburg PA
CBHW031813220426
43662CB00007B/631